Data Communication

97006470

Second Edition

Addison Wesley Longman Limited
Edinburgh Gate, Harlow
Essex CM20 2JE, England
and Associated Companies throughout the world.

First published 1991
Second edition 1995
Reprinted 1996 (Twice)

British Library Cataloguing in Publication Data
A catalogue entry for this title is available from the British Library.

ISBN 0-582-24520-6

Library of Congress Cataloging-in-Publication Data
A catalog entry for this title is available from the Library of Congress.

Produced through Longman Malaysia, PP

Contents

Preface

This book has been written to provide a comprehensive introduction to data communications for students of the Business and Technical Education Council (BTEC) Certificate/Diploma courses in Electrical, Electronic, Communications, and Computer Engineering. The book provides full coverage of the unit Data Communications and of the network and communications parts of the unit Information Handling Systems. An attempt has been made throughout to keep the treatment of each topic to a level appropriate to a Level III BTEC student. It has been assumed that the reader will possess a prior knowledge of electrical principles, electronics, and, in particular, microprocessors at the BTEC Level II/III standard.

In Chapter 4, which deals with interfaces and interface chips, it has been necessary to consider particular devices and I have chosen the Motorola MC6821 PIA and the Motorola MC 6850 ACIA. Although these devices may not often be used in new designs they are still in common use in both industry and education, as enquiries from some component distributors and colleges have shown, and they are relatively simple in their operation.

I wish to express my thanks to Motorola Ltd for their permission to reproduce some diagrams from their data sheets.

DCG

1 Introduction to Data Communication

The term 'data' throughout this text means information that is represented by a binary electrical signal. Data transmission means the transmission of data between two computers, or between a computer and a terminal. The Consultative Committee International Telephony and Telegraphy (CCITT), now known as the International Telecommunications Union—Telephony (ITU-T), call terminals data terminal equipment (DTE) but in this book the word terminal will be employed. The computers in a data network may consist of one, or more, mainframe or host computers, minicomputers, and micro-, or personal computers. A wide variety of peripherals are in common use and they include such equipments as disc drives, printers, plotters, visual display units (VDUs) and keyboards. Besides being able to communicate with local terminals, and/or peripherals, a computer must also be able to communicate with other computers and/or terminals that are installed at more distant locations.

The applications of data communications are many, widespread, and continually increasing in number. The Government, large corporations and financial institutions, such as banks and building societies, have installed extensive data communications networks to rapidly transfer large amounts of data from one point to another, to calculate wages and salaries and to print out pay-slips, etc., and to produce bills and invoices. The majority of the work carried out by a computer in such a network consists of the processing and storage of data and then the outputting of this data at regular intervals, say weekly or monthly. The flow of such tasks can easily be planned in advance; the data is prepared and processed in batches and hence this type of operation is known as *batch processing*. Other early applications of data communications were restricted to such systems as airline seat reservations, package holiday bookings, remote job entry (i.e. a user transmits data over a link to a computer, the computer carries out whatever tasks are required and then transmits the results back down the link for printout, or use, at the entry terminal) and data acquisition, and these had little direct effect upon everyday life. In recent years, however, further applications of data communications have become commonplace and these include: (*a*) on-line credit card checking, (*b*) electronic fund transfer for moving cash from one bank to another, (*c*) bank and building society wall cash dispensers, known as automatic telling machines (ATM), (*d*) electronic point-of-sale systems where people can buy goods and have their bank accounts automatically debited, (*e*) electronic mail, (*f*) video text systems such as Prestel, (*g*) facsimile telegraphy (FAX) systems, and (*h*) transaction processing. A further modern application, which is frequently seen on the television news, is in the stock exchange; dealers selling and buying shares, or currency, of enormous monetary value, rely upon their ability to communicate reliably, rapidly, and accurately with other dealers around the world.

The important constraints on all such data communication systems are (*a*) the response time, (*b*) the throughput, and (*c*) the human factor. The response time is a measure of the speed of operation of the system. In many systems a fast response time is essential; think, for example, of a wall cash dispenser. When a user taps in his PIN number and requests a cash sum he expects to receive his money from the machine within a fairly short time. The throughput is a measure of the load on a system, i.e. the percentage time a link is occupied for a given number of messages sent over that link. The throughput of a system should be as high as possible so that the maximum utilization of expensive lines and terminal equipment is achieved. The terminals used should be as easy to operate as possible both to reduce the possibility of human error and to increase the operating speed. Clearly, this human factor is of more importance in situations where the terminal will be used by an untrained person; once again the wall cash dispenser can be quoted as an example.

Many computer networks consist of a host computer which is often a mainframe computer and which is linked to a number of terminals and also, perhaps, to some other computers. The terminal can be located in the near vicinity of the host computer or it may be located many miles away, perhaps in another country or even in another continent. For communication with nearby terminals a *local area network* or LAN can be used. For data communication between more distant terminals either a dedicated circuit or a switched circuit, via either the public switched telephone network (PSTN), or the public switched data network (PSDN), or a private network must be used.

In a dedicated network every terminal has a connection to the host computer that is maintained whether or not data is being transmitted. In a switched network a physical connection between the host computer and a terminal is only set up when a message is to be passed. Once the message has been received the connection is cleared down. Clearly, a switched network is able to support a much larger number of terminals than can a dedicated network. Two switched networks are available to many users in the UK; one, known as the public switched telephone network (PSTN) was originally designed for the transmission of speech signals. Long-distance data communications, dedicated or switched, may be routed, wholly or partly, over some combination of terrestrial link via coaxial or fibre optic cable, or microwave radio, or may be routed over a communication satellite system.

Local area networks (LANs), which connect together a number of micro-computers, (generally known as personal computers or PCs) and a number of peripherals, such as printers, are now used by all large organizations. Since such organizations, be they official or private, operate from many different locations, the need has developed for the inter-connection of LANs. Data communication networks are therefore also employed to inter-connect LANs. High-speed LAN inter-connections now replace many of the earlier terminal-to-host computer networks.

The core network of the UK telephone system uses digital techniques but the access network is still analogue in nature. When a connection via the PSTN or an analogue leased line is used to transmit data signals it is necessary to convert the data signals from their digital form into the corresponding voice-frequency signals. This conversion is carried out by a *modem*.

Also available, certainly at many locations throughout the UK, is the public switched data network (PSDN); this is called *Switchstream* by BT. Switchstream provides a high-speed data switching network which conforms with the ITU-T standard X25. A user can set up a connection between a terminal and a host computer as and when required and then use this connection for the transmission of data. The use of the X25 protocol provides the following facilities to the user: (*a*) error checking, (*b*) unique network addresses, (*c*) speed conversion, so that a user is unaware of any speed difference between the calling and the called ends of the connection. Since the X25 protocol is internationally accepted it is possible to use the network to make calls to most other countries.

The use of the PSDN is generally advantageous when several users wish simultaneously to transmit low volumes of data traffic over long distances — particularly at peak hours. It may, however, work out to be more expensive for the occasional user.

Other fully digital services are also made available by BT, and are known as *Kilostream*, *Megastream* and *Satstream*, which are all eminently suitable for the transmission of data signals. Kilostream and Megastream are both point-to-point digital services; Kilostream can operate at speeds of 2400 bits/s, 4800 bits/s, 9600 bits/s, 48 kbits/s, and 64 kbits/s, Megastream can operate at 2.048 Mbits/s, 8 Mbits/s, 34 Mbits/s and 140 Mbits/s.

Serial and Parallel Transmission of Data Signals

The characters in a computer system are represented by a data code each element of which consists of a group of binary digits, or *bits*. Each bit can only be either binary 1 or binary 0. The movement,

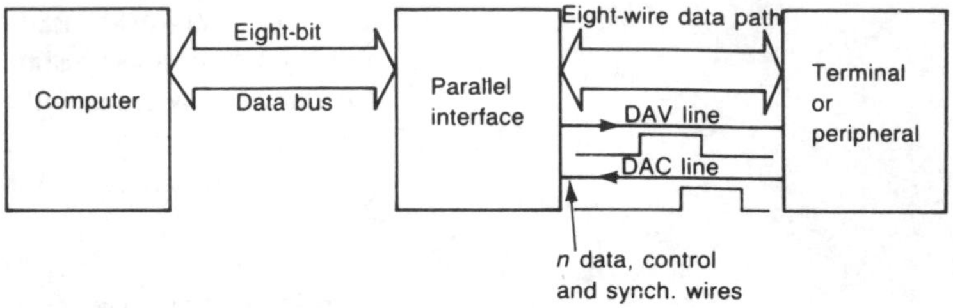

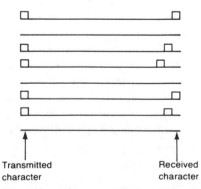

Fig. 1.1 Parallel transmission of data.

storage and processing of data within a computer, or a microprocessor, may be carried out on a 16-bit, or 32-bit basis, depending upon the type of computer. A group of eight bits is known as a *byte*. Outside of the computer, or the microprocessor, data can be transmitted to a peripheral, a terminal, or a modem using either parallel or serial transmission.

With the parallel transmission of data all the bits making up a character are transmitted simultaneously over separate conductors, as shown by Fig. 1.1. When a computer has data to send to a terminal it takes the data-available (DAV) line high. When the terminal is ready to receive the data it will take the data-accepted line (DAC) line high. This *handshaking* procedure occurs for each character in the message transmitted by the computer. Handshaking is used to provide more precise timing of the data transfer between a computer and a terminal or a peripheral. Handshaking of some kind is generally necessary because a computer and a communicating terminal will almost certainly operate at different speeds. Extra handshake lines are usually employed to control the timing of the data transfer and this is discussed in Chapter 4. The number of conductors required for a parallel interface is known as the *bus width*; in the figure it is 10. Each of the conductors has a dedicated function: some carry data and others carry control and synchronization information. Since several conductors are necessary the parallel transmission system is only economic over fairly short distances.

The cost of a multi-conductor cable is relatively high and a problem known as *skew* may occur. Skew is an effect in which the bits transmitted simultaneously into a number of different conductors in the same cable may arrive at the far end at different times. This effect is illustrated by Fig. 1.2. The effect worsens with increase in the length of the cable and it will very soon lead to errors in the received data. A parallel interface would give fast data transfer and would be relatively simple to operate but the bus width is at least 10 and usually higher since one, or more, extra conductors, known as the handshake line(s), are also necessary. Parallel transmission is used between a computer and high-speed printers and disc drives since a high-speed connection is wanted and the cable length is short.

For all but the shortest peripheral connections serial data transmission is normally employed and Fig. 1.3 illustrates the basic concept. The internal parallel data is applied to a parallel-to-serial

Fig. 1.2 Showing the effect of skew on parallel data.

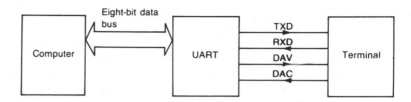

Fig. 1.3 Serial transmission of data.

converter and here it is changed into serial form. The parallel-to-serial converter is usually within an IC which also performs some other functions and is known variously as a UART, a VART, an ACIA, a PIA, etc. The serial port transmits each data character on an element-by-element basis and so only two conductors, transmit data (TXD) and receive data (RXD), are required. Each signalling element may be equal to one bit, equal to two bits (known as dibits), three bits (tribits), four bits (quabits), and five bits (quinbits), or even less than one bit (Manchester encoding). In this chapter an element will always be assumed to be equal to a bit. Since the bits are transmitted sequentially and not simultaneously the *data transfer rate* is considerably less than for parallel transmission. The least significant bit is transmitted first and the most significant bit last. Each transmitted character is represented by a particular sequence of bits according to the code employed. The receiver must sample the incoming data signal at the correct instants in time before it can reproduce the transmitted character.

Serial transmission gives rise to three synchronization problems: namely bit synchronization, character synchronization and block synchronization. Suppose that the serial data signal is 00100100. For correct reception the time intervals used by the transmitter and the receiver must be equal to one another. To achieve this both the transmitter and the receiver include *clocks*, the term clock is used for any source of timing pulses. The receiver clock is required to indicate the precise instants in time at which the serial data signal must be sampled by the receiver to determine the logical state of each received bit. The receiver's clock pulses should, as shown by Fig. 1.4, ideally occur at the middle of each time period occupied by a received bit. If the clock frequency were to be reduced to one-half that shown in the figure the incoming data signal would be read by

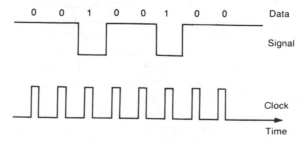

Fig. 1.4 Clocking serial data.

the receiver as either 0100, or as 0010, depending upon the timing of the first clock pulse. Thus it is necessary, for correct data reception, for the receiver's clock to be in synchronism with the transmitter's clock, see page 7. When the receiver has attained bit synchronization it must then achieve character synchronization, this requires that the receiver must be able to decide which group of bits belong to which character. Essentially, this means that the receiver must be able to determine which bit is the first bit (least significant bit) of a character. It is also necessary that the receiver is able to recognize the beginning and the end of each block of data. The necessary synchronization can be achieved either *synchronously* or *non-synchronously*. Data transmitted by the terminal to the computer over the RXD line is applied to a serial-to-parallel converter, which is also within the UART (or other device) IC, and converted to parallel form before it is passed on to the computer.

Non-synchronous Data Transmission

With non-synchronous transmission of data each character is transmitted as an independent entity; this means that the time between the last bit of one character and the first bit of the following character is not fixed. Non-synchronous data transmission is simpler than synchronous transmission because only the data signal is transmitted over the line. The receiver clock is generated locally within the receiver and it is kept in synchronism with the transmitter clock by the use of *start* and *stop* bits which are transmitted with every character. In the idle condition the line voltage is held at the binary 1 level by the transmitter, or repeated 1 bits are transmitted, and the receiver clock is stopped. When the transmitter has a character to send it first switches the line voltage to the binary 0 level for a one-bit period of time, known as the start bit, and then it sends the bits that represent the character. The receiver's clock is started by the transition from 1 to 0 of the leading edge of the start bit and it then runs freely to generate the clock pulses. The first clock pulse should occur after a time interval of approximately 1.5 bits and each subsequent bit should be sampled at one-bit time intervals thereafter. This means that the receiver clock is usually synchronized to ensure that the clock transitions occur about half-way through the time period of each bit received. The sampling of each bit will then occur near its centre, which is desirable since it minimizes the probability of error. At the end of each character a *stop* bit, which is at the binary 1 voltage level, is transmitted to stop the clock in the receiver. The receiver clock then waits until the next start bit arrives. The synchronization between the transmitter and the receiver clocks is carried out on a character-by-character basis. This means that the receiver clock need not be very stable. If, for example, the timing is to be allowed to drift by up to ± 0.2 of a bit period, at the end of a 10-bit character the receiver's clock must be stable to within $\pm 100/(10 \times 5) = \pm 2\%$

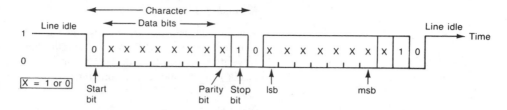

Fig. 1.5 Start–stop synchronization.

of the clock in the transmitter. This is easily achieved if either a crystal oscillator or a phase-locked loop is employed in the clock circuitry.

The waveform of a data signal using start–stop synchronization is shown by Fig. 1.5. The start and stop bits do not themselves carry any information but merely indicate the beginning and end of each character. It can be seen that an eighth bit, labelled as a *parity bit*, is included in the waveform; this is a bit that is set to either binary 1 or 0 to ensure that the total number of 1s in a character is always either even for even parity, or odd for odd parity. Thus each character has a length of ten bits. This parity system permits the detection of single errors in a character.

The efficiency of the non-synchronous system is not very high since only seven of the 10 bits that are transmitted to represent a character actually contain information. The term anisochronous is sometimes applied to describe a channel that is capable of transmitting but not timing signals.

If the receiver clock runs at a different speed from the transmitter clock it will not be long before a received bit is missed by the receiver. If the clock in the receiver runs slightly faster than the clock in the transmitter the receiver will sample the incoming data stream earlier and earlier in each bit period. After some short time the receiver will sample the same bit twice, and from then on the received data will be out of synchronism with the transmitted data. Suppose the duration of each received bit is x ms and the sampling period of the receiver (i.e. 1/(sampling frequency)) is y ms. Then, see Fig. 1.6,

$$nx + x/2 = ny \tag{1.1}$$

is the point at which an error will occur because the samples will miss a received bit. This is illustrated by Example 1.1.

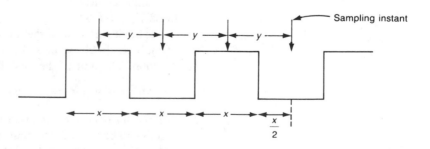

Fig. 1.6

Example 1.1

A non-synchronous signal is transmitted at 1200 bits/s. If the clock in the receiver is running (*a*) 2% , and (*b*) 1% slow how many bits will be received correctly before an error occurs? Assume that both the transmitter and the receiver clocks start at the same time.

Solution

(*a*) The clock in the receiver runs at $1200 - 24 = 1176$ bits/s.
Therefore $x = 1/1200$ and $y = 1/1176$.

$n/1200 + 1/2400 = n/1176$
$1/2400 = n(1/1176 - 1/1200) = n(1.7 \times 10^{-5})$
or $n = 1/(2400 \times 1.7 \times 10^{-5}) = 25$ bits.   (*Ans.*)

(*b*) 1% slow $= 1188$ bits/s and the error occurs after 50 bits have been received.   (*Ans.*)

Most interface chips, such as an ACIA or a UART (p. 83) make a majority judgement of data on a basis of several samples. This makes modems with a bias distortion greater than 25% unstable. A typical bias distortion figure is about 3%.

Non-synchronous transmission of data signals is often employed since it is both relatively simple and inexpensive. However, it is only suitable for low-speed data circuits for two reasons: (*a*) its reduced data transmission efficiency, and (*b*) a free-running receiver clock only provides satisfactory synchronization at fairly low speeds. It is, however, often used for links between a computer and a nearby terminal at speeds of up to about 19 kbits/s.

Data Transmission Rates

The data signalling rate is the speed with which information can be transmitted over a circuit and it is expressed in bits/s. For serial transmission this is equal to

$$\text{Data signalling rate} = \log_2 n/T \text{ bits/s,} \tag{1.2}$$

where n is the number of possible bit combinations, and T is the duration of each line signal element. If there are only two possible bit combinations, 1 and 0, $n = 2$ then the data signalling rate is $\log_2 2/T = 1/T$ bits/s.

The modulation rate is the rate at which changes in the logical state of the line circuit occur and it is equal to the reciprocal of the bit duration. The unit of the modulation rate is the baud, i.e.

$$\text{Modulation rate} = 1/T \text{ bauds.} \tag{1.3}$$

Clearly, the data signalling rate and the modulation rate (or the baud speed) will be equal to one another if there are only two possible line states. If, however, the data bits are grouped into twos — to form the *dibits* 00, 10, 11, and 10 — before transmission to line there will be four possible line states and then the data signalling rate will be

$$\text{Data signalling rate} = \log_2 4/T = 2/T \text{ bits/s}, \qquad (1.4)$$

i.e. twice as high as the baud speed.

Example 1.2

A data circuit has a transmission speed or modulation rate of 2400 bauds. Determine the possible bit rate if the data stream is encoded into (a) tribits, and (b) quabits.

Solution

(a) With the tribits 000, 001, 010, etc. there are 8 possible bit combinations. Hence $n = 8$.

$$\text{data signalling rate} = (\log_2 8)/T = 3/T$$
$$= 3 \times 2400 = 7200 \text{ bits/s} \qquad (Ans.)$$

(b) With quabits 0000, 0001 etc. $n = 16$ and,

$$\text{data signalling rate} = (\log_2 16)/T = 4/T$$
$$= 4 \times 2400 = 9600 \text{ bits/s} \qquad (Ans.)$$

For parallel transmission the data transmission rate is the sum of the bit rates on each of the m conductors, i.e.

$$\text{Data signalling rate} = m\log_2 n/T, \qquad (1.5)$$

where m is the number of conductors and n and T are as before.

Bias Distortion

When a digital signal is transmitted over a telephone circuit the characteristics of that circuit will cause the received signal to be both reduced in amplitude and distorted; in addition noise voltages and interfering signals will be picked up to still further distort the signal. The combined effects will result in the received signal being subjected to both distortion and jitter. Figure 1.7 shows the effects of bias distortion on a data signal. The received signal is sampled at the receiver at regular intervals of time to determine whether the signal level at that moment is at the binary 1 or the binary 0 level. If the sampling threshold is set at too high a level the 1 bits will be lengthened and the 0 bits will be shortened. If the threshold is too low the 1 bits are shortened and the 0 bits are lengthened. In either case the correct mark/space ratio will be lost and if the distortion is excessive it will result in some of the bits being incorrectly received. The term positive bias refers to the binary 1 pulses being lengthened, and negative bias to the 0 pulses becoming longer.

The percentage bias distortion $= 100(T_1 - T_2)/[2(T_1 + T_2)]\%$,
$$(1.6)$$

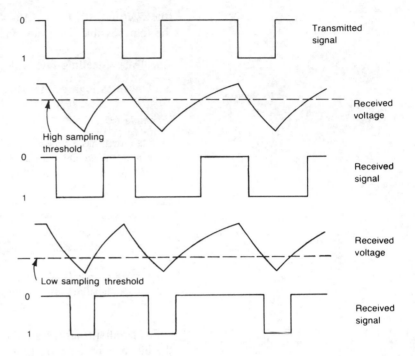

Fig. 1.7 Bias distortion.

where T_1 and T_2 are the time durations of the binary 1 and the binary 0 pulses, respectively.

Bit Error

The data transitions in the received data waveform tend to move around from their ideal positions in time. This results in an effect that is known as *bit jitter* and it is illustrated by Fig. 1.8. If τ is the duration of a pulse and t is the movement of a pulse from its ideal position then

$$\text{Bit jitter} = t_{max} - t_{min}. \qquad (1.7)$$

When it is expressed as a percentage of the pulse duration τ

$$\text{Bit jitter} = (t_{max} - t_{min})/\tau \times 100\%. \qquad (1.8)$$

A typical figure for pulse jitter is $100\,\mu s$, giving a percentage figure that depends upon the bit rate.

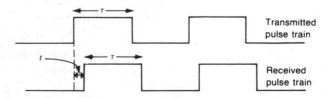

Fig. 1.8 Bit jitter.

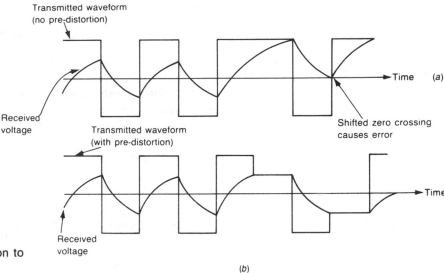

Fig. 1.9 Use of pre-distortion to reduce error.

At the higher bit rates used in LANs (p. 159) several bit periods may overcharge the line capacitance and produce a shift in the point at which the received signal passes through zero volts. This, in turn, will probably cause errors in the recovery of the data signal. This effect is shown by Fig. 1.9(a). The effect can be overcome by the use of pre-distortion as shown by Fig. 1.9(b), in which the transmitted signal is reduced to the level consistent with a short pulse for the second half of each long pulse. This results in the received signal crossing the zero voltage axis at approximately the same point regardless of the pulse length.

Bit Error Rate

Any data circuit is always subjected to noise and interference voltages that originate from a wide variety of sources. These unwanted voltages are superimposed upon the received data voltage and they will corrupt the waveform. At each sampling instant the receiver must determine whether the bit received at that moment is a 1 or a 0 and any waveform corruption increases the probability of this determination being incorrect and hence of an error occurring. The *bit error rate* (BER) is given by

$$BER = \frac{\text{number of bits wrongly received}}{\text{total number of bits transmitted}} \qquad (1.9)$$

Example 1.3

A message is transmitted at 2400 bits/s and it occupies a time period of 1 minute and 20 seconds. If two of the received bits are in error calculate the BER.

Solution
At 2400 bits/s there will be no start or stop bits and so the total number of bits transmitted is $80 \times 2400 = 192\,000$. Therefore

$$\text{BER} = 2/192\,000 = 10.42 \times 10^{-6}. \quad (\textit{Ans.})$$

The BER depends upon the *signal-to-noise ratio* of the circuit employed to carry the data signal. Signal-to-noise ratio is the ratio of the wanted signal power to the unwanted noise power, i.e.

$$\text{Signal-to-noise ratio} = \frac{\text{wanted signal power}}{\text{unwanted noise power}}, \quad (1.10)$$

or, quoted in decibels,

$$\text{Signal-to-noise ratio} = 10 \log_{10} \frac{(\text{wanted signal power})}{(\text{unwanted noise power})}. \quad (1.11)$$

The minimum signal-to-noise ratio required for a particular circuit depends upon the highest BER which is acceptable since there is a clear relationship between the two quantities.

The various sources of noise that can affect a data communication circuit are, briefly, as follows.

(*a*) thermal agitation noise in conductors, resistors and semiconductors;
(*b*) shot noise and flicker noise in semiconductors;
(*c*) faulty electrical connections which may cause short breaks in the transmission path;
(*d*) electrical and magnetic couplings to other circuits, causing cross-talk in equipment wiring and in cables, etc.

Impulsive noise, i.e. noise that appears for short periods of time and which is therefore not continuous, may often be of large amplitude relative to the signal level. This form of noise is at its most troublesome when the attenuation of the line is fairly high and when the bit rate is high. Suppose, for example, that a burst of noise lasts for 0.1 ms; if the bit rate is 1200 bits/s each bit lasts for 0.83 ms and it is unlikely that an error will occur. If the bit rate is 4800 bits/s each bit lasts for 0.2 ms and the probability of an error is high. This means, of course, that the error probability increases with increase in the bit rate.

It might seem that an improved signal-to-noise ratio, and hence a lower BER, could always be obtained by increasing the transmitted signal power. This, however, increases the level of crosstalk into neighbouring circuits and so the transmitted signal power is limited by the ITU-T recommendation V2. In the UK modems are set to have a transmitted power level such that a level of -10 dBm is produced at the nearest digital trunk exchange.

Synchronous Data Transmission

With a synchronous data system blocks of data are transmitted continuously *without* either start or stop bits. The clock in the receiver

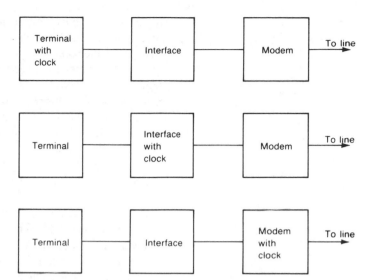

Fig. 1.10 In a synchronous system the clock may be in (*a*) the terminal, (*b*) the interface, or (*c*) the modem.

runs continuously and it is locked into synchronism with the clock in the transmitter. To achieve the necessary synchronization clocking information must be transmitted over the circuit along with the data, perhaps by the use of some method of encoding the data so that it includes the clocking information, or by the use of a modem (p. 36) that encodes the clocking information during the modulation process, or by having the clock in the interface circuitry. These three methods are outlined by Fig. 1.10. In all cases, the data is transmitted at a defined rate which is controlled by the transmitter clock. Obviously, the receiver must include circuitry which enables it to decode the received clock information. Since the transmitter clock determines both the send and the receive rates, synchronism is maintained. A channel that is able to transmit the clocking information in addition to the data is known as a synchronous or *isochronous* channel.

Bytes of data are continuously transmitted without any gaps for as long as is necessary for the message to be sent. The time interval between the last bit of one character and the first bit of the next character is either zero or it is an integer multiple of the time period occupied by a character. If any gaps do occur in the bit stream the transmitter will insert padding bytes to fill the gaps. There is no requirement for start and stop bits to be employed. Figure 1.11 shows a synchronous bit stream.

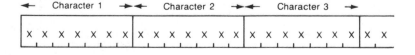

Fig. 1.11 Synchronous bit stream.

The receiver must commence sampling the incoming data at the middle of the first bit of the first character; otherwise it will be out of step with the received signal and it will receive it incorrectly. Once bit synchronization has been achieved the receiver must know which group of received bits constitutes a character (character synchronization). This requires that the receiver monitors the received data on a bit-by-bit basis until it recognizes a character synchronization pattern. This will then allow the receiver to determine which set of received bits defines the first transmitted character. Subsequent characters can then be recognized by the receiver. The principle of synchronous transmission will be illustrated by the BiSynch protocol that is discussed in greater detail in Chapter 8.

The data stream is usually preceded by two, or more, synchronization (SYN) bytes which have a particular pattern that the receiver can recognize. The synchronous receiver is initially in its search mode and it looks for two successive SYN characters in the incoming bit stream. Once two such bytes have been identified the data is moved into a temporary store (a shift register) and a character-available flag is raised after every eight bits. Thus, a short message will be of the form shown by Fig. 1.12; the transmission commences with two SYN characters and then a STX character. This is a start-of-text control character that indicates the start of the actual data information. Once the first bit of the first data character has been identified the receiver keeps a count of the following bytes, converts them to characters and so assembles the complete message. ETX indicates the end-of-text and FCS indicates a frame-check sequence, this is a 16-bit signal that contains error-checking information. Some kind of *header* information may precede the message itself. This contains addressing and/or administration information, e.g. the destination address, the originating address, priority information, and/or the date and time of transmission. If a header is used a SOH character is transmitted to inform the receiver that the following information is header information. The STX character then indicates the start of the actual message. Figure 1.13 shows the idea. The combination of the data plus the control information is called a *frame* and its format depends upon whether it is bit orientated or character orientated; the BiSynch system that has been used as an example here is an example of a character-orientated protocol.

Fig. 1.12 Synchronous message format.

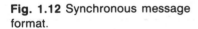

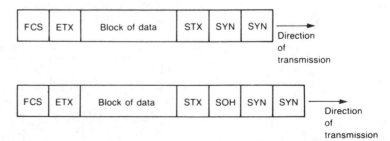

Fig. 1.13 Message with header.

FCS	ETX	Block of data 3	STX	ETB	Block of data 2	STX	ETB	Block of data 1	STX	SOH	SYN	SYN

Direction
of transmission

Fig. 1.14 Blocks of data.

A long message will often be split into a number of *blocks*. Each block is preceded by the STX character and is followed by the ETB character, the final block is followed by the ETX character; see Fig. 1.14. As before, the ETX character denotes that the complete message has been received.

Synchronous versus Non-synchronous Working

Both non-synchronous and synchronous transmission methods are commonly employed by data terminals. Low-speed terminals are always non-synchronous while many VDUs and remote job-entry batch terminals are synchronous. Personal computers nearly always employ non-synchronous transmission over telephone circuits and very often on their printer interfaces as well. The choice between the two methods must be based upon the required speed of response, and telephone circuit costs.

Non-synchronous transmission is comparatively inexpensive. Bytes arrive at the receiver separated by start and stop bits which makes character synchronism easy to achieve. Since the clock in the receiver is started anew for each received character and it only needs to stay in synchronism for an eight-bit period of time, bit synchronism is also no problem. Non-synchronous transmission is only suitable for low-speed data because: (*a*) the start/stop bits give an overhead of two bits in every ten to reduce the maximum data transfer efficiency to 80%; and (*b*) a free-running clock is only satisfactory at low bit rates.

Synchronous transmission is more expensive than non-synchronous operation but it is able to work at much higher bit rates. Because the data is typically sent with zero gap between consecutive characters some level of buffering is needed both at the transmitter and at the receiver. The transmission speed can be altered by changing the transmitter clock and the data rate at the same time. Finally, the overhead is much smaller than for non-synchronous operation, typically about 5%.

Non-synchronous channels are often used for 'within-house' terminal-to-terminal communications. With personal computers the serial port is non-synchronous and it can be used to connect a modem or a printer to the computer, or perhaps to connect the personal computer to a minicomputer so that it can operate as an intelligent terminal. Synchronous channels are used in many mainframe, or host, computer data networks since a greater *throughput* is required in an environment where many terminals are connected to the host computer and are multiplexed onto a single channel. Although most synchronous terminals are used in conjunction with synchronous (or isochronous) channels some such terminals can also operate over anisochronous channels. Similarly, non-synchronous terminals can operate over either anisochronous or isochronous channels.

Precise definitions of the terms used in conjunction with synchronous transmission are as follows. (*a*) *Synchronous*: a signal is synchronous if its corresponding significant points occur at precisely the same average rate. (*b*) *Isochronous*: a signal is isochronous if the time interval between any two significant points is equal to the unit interval, or a multiple of the unit interval.

Data Codes

In a data communication system characters must be transmitted over communication channels from one point to another. The characters cannot be transmitted directly as they stand but must first be coded using a recognized code. The vast majority of data terminals are designed to use either one of two internationally accepted codes:

(*a*) The International Standards Organization (ISO) seven-bit code which is known as the International Alphabet No. 5 (IA5); the United States version of this — the American Standard Code for Information Interchange (ASCII) is nearly always used and it is given in Table 1.1. ASCII can give $2^7 = 128$ different binary numbers, 32 of which are reserved for control functions such as SYN and STX. This is enough to allow a full upper-case, and lower-case alphanumeric set plus additional characters such as =,/ and ?.

(*b*) The Extended Binary-coded Decimal Interchange Code (EBCDIC) is an eight-bit code that is used by some IBM terminals.

Table 1.1 ASCII character codes

Character	Binary	Hex	Decimal	Character	Binary	Hex	Decimal
Null	00000000	00	000	@	01000000	40	64
SOH	00000001	01	001	A	01000001	41	65
STX	00000010	02	002	B	01000010	42	66
ETX	00000011	03	003	C	01000011	43	67
EOT	00000100	04	004	D	01000100	44	68
ENQ	00000101	05	005	E	01000101	45	69
ACK	00000110	06	006	F	01000110	46	70
BEL	00000111	07	007	G	01000111	47	71
BS	00001000	08	008	H	01001000	48	72
HT	00001001	09	009	I	01001001	49	73
LF	00001010	0A	010	J	01001010	4A	74
VT	00001011	0B	011	K	01001011	4B	75
FF	00001100	0C	012	L	01001100	4C	76
CR	00001101	0D	013	M	01001101	4D	77
SO	00001110	0E	014	N	01001110	4E	78
SI	00001111	0F	015	O	01001111	4F	79

Table 1.1 (Cont'd)

Character	Binary	Hex	Decimal	Character	Binary	Hex	Decimal
DLE	00010000	10	016	P	01010000	50	80
DC1	00010001	11	017	Q	01010001	51	81
DC2	00010010	12	018	R	01010010	52	82
DC3	00010011	13	019	S	01010011	53	83
DC4	00010100	14	020	T	01010100	54	84
NAK	00010101	15	021	U	01010101	55	85
SYN	00010110	16	022	V	01010110	56	86
ETB	00010111	17	023	W	01010111	57	87
CAN	00011000	18	024	X	01011000	58	88
EM	00011001	19	025	Y	01011001	59	89
SUB	00011010	1A	026	Z	01011010	5A	90
ESCAPE	00011011	1B	027	[	01011011	5B	91
FS	00011100	1C	028	bkslh	01011100	5C	92
GS	00011101	1D	029	]	01011101	5D	93
RS	00011110	1E	030	|	01011110	5E	94
US	00011111	1F	031	←	01011111	5F	95
SPACE	00100000	20	032	-	01100000	60	96
!	00100001	21	033	a	01100001	61	97
"	00100010	22	034	b	01100010	62	98
#	00100011	23	035	c	01100011	63	99
$	00100100	24	036	d	01100100	64	100
%	00100101	25	037	e	01100101	65	101
&	00100110	26	038	f	01100110	66	102
'	00100111	27	039	g	01100111	67	103
(	00101000	28	040	h	01101000	68	104
)	00101001	29	041	i	01101001	69	105
*	00101010	2A	042	j	01101010	6A	106
+	00101011	2B	043	k	01101011	6B	107
,	00101100	2C	044	l	01101100	6C	108
—	00101101	2D	045	m	01101101	6D	109
.	00101110	2E	046	n	01101110	6E	110
/	00101111	2F	047	o	01101111	6F	111
0	00110000	30	048	p	01110000	70	112
1	00110001	31	049	q	01110001	71	113
2	00110010	32	050	r	01110010	72	114
3	00110011	33	051	s	01110011	73	115
4	00110100	34	052	t	01110100	74	116
5	00110101	35	053	u	01110101	75	117
6	00110110	36	054	v	01110110	76	118
7	00110111	37	055	w	01110111	77	119
8	00111000	38	056	x	01111000	78	120
9	00111001	39	057	y	01111001	79	121
:	00111010	3A	058	z	01111010	7A	122
;	00111011	3B	059	{	01111011	7B	123
<	0011100	3C	060	--	01111100	7C	124
=	00111101	3D	061	}	01111101	7D	125
>	00111110	3E	062	~	01111110	7E	126
?	00111111	3F	063	DEL	01111111	7F	127

In the UK the £ sign is required instead of the $ sign. When the ASCII code is used for the serial transmission of data eight bits are sent per character with the eighth bit being used as a *parity bit*. The use of parity bits as a means of locating errors in the received data will be discussed in Chapter 9. Briefly, each character has one extra bit added. The extra bit may be either a 1 or a 0; the choice being made so that the total number of 1 bits transmitted is odd if *odd parity* is used and even if *even parity* is employed. The receiver checks each received character to determine how many 1 bits it contains and if the total is odd (or even) the received character is taken as being correct. If it is not an error has occurred which must be corrected. If parity checking is not employed the eighth bit is set permanently to be either 1 or 0. The least significant bit (lsb) is the right-hand bit and it is transmitted to the line first, the parity bit is the left-hand bit and it is transmitted last. In the table the parity bit is shown as a 0 throughout.

There are a total of 32 control characters in Table 1.1. Most of these fall into one of four groups: (*a*) device controls, (*b*) format controls, (*c*) information separators, and (*d*) transmission controls.

Device Controls

The four device controls are DC1, DC2, DC3 and DC4 and they can be used to control the physical operation of a terminal, e.g. to turn motors on or off. The actual usage is determined by the manufacturer of the terminal. Often DC1 and DC3 are used to control the flow of data to a non-synchronous terminal, DC1 turning the data flow on and DC3 turning it off.

Format Controls

The six format controls are:

(*a*) BS or back space, which makes a printer head, or a VDU cursor, move one position back;

(*b*) HT or horizontal tabulation, which causes a printer head, or VDU cursor, to move horizontally a predetermined amount;

(*c*) LF or line feed, which causes the printer head, or cursor, to move to the same character position on the next line;

(*d*) VT or vertical tabulation, which moves the printer head, or VDU cursor, to the same character position a predetermined number of lines further on;

(*e*) FF or form feed, which moves the printer head, or VDU cursor, to the same character position on a predetermined line on another page; and (*f*) CR or carriage return, which causes the printer head, or VDU cursor, to move to the first position on the next line.

Information Separators

The four information separators are used to separate transmitted information to make it easier to handle records, etc. They are:

(*a*) US or unit separator which is used to separate units of data;

(*b*) RS or record separator, which is used to separate a number of units, or a record, of data;

(*c*) GS or group separator is used to separate a number of records, or a group; and

(*d*) FS or file separator which is used to separate files of information where a file is a number of groups.

Transmission Controls

The transmission control characters are used to frame messages into recognized formats and also to control the flow of data in a network. They are used in conjunction with character-orientated protocols (Chapter 8).

There are also some other control characters that do not fit into any of the above groups. These are:

(*a*) BEL, which is used to attract the attention of a human being (it may actually ring a bell);

(*b*) SO (shift out) indicates that the characters that follow should be interpreted by the receiver as being outside of the standard ASCII set of characters until the SI (shift in) character is received;

(*c*) CAN (the cancel character) indicates to the receiver that the data that has preceded it should be disregarded;

(*d*) EM (the end-of-medium character) indicates that the end of a tape, or some other physical medium, has been reached; and

(*e*) DEL (the delete character) is used to delete any unwanted characters.

Full-duplex and Half-duplex

Nearly all data circuits are operated on either a half-duplex or on a full-duplex basis. A half-duplex circuit is able to transmit data in either direction but only in one direction at any one time. The process of changing the direction of transmission requires extra computer software and it occupies a certain amount of time that is known as the turnaround time. In some cases this can amount to several milliseconds; when it occurs frequently the turnaround delay can significantly reduce the throughput of the circuit.

A full-duplex circuit is one that is able to transmit data in both directions at the same time. In many cases separate channels are used for each direction of transmission. Very often full-duplex operation of a circuit is employed even though there is no requirement for the simultaneous transmission of data in both directions. This is done because otherwise the turnaround time would lead to unacceptably long response times from the computer. Networks that employ cheap mini-, or micro- computers and/or simple terminals also often use full-duplex operation to keep the terminal costs down.

Standards

All data communications, but particularly international data communications, have always faced problems with incompatible standards. Initially, standards were set by the equipment manufacturers, particularly of mainframe, or host, computers, but this practice restricted a customer to a single manufacturer for all his computing and communications equipment. Nowadays, most aspects of data communication are covered by international standards based upon the recommendations of the ITU-T. The ITU-T recommendations for data communications over the telephone network are given by the V-series and include specifications for modems, interfaces, test equipment and line quality. They are shown by Table 1.2.

Data Terminals

A data terminal equipment (DTE) is any equipment that is able to transmit and/or receive data signals. This definition includes both intelligent and non-intelligent, or dumb, equipments. The term intelligent means that the equipment has the capability to carry out some computing itself and it may actually be a minicomputer or a micro-(personal) computer. Intelligent terminals may employ either the ASCII code or the EBDIC code but dumb terminals always use the ASCII code. The requirements of a terminal may vary enormously with its application(s); some memory will be necessary if data storage is required, or if the speed of transmission is to be changed either up or down to suit the characteristics of the transmission medium. If, for example, the input signal to the terminal is provided by a human keyboard operator the input data rate will be very low — probably less than five characters per second (or about 40 bits/s) and it will certainly need to be considerably increased before it is transmitted to line.

Conversely, some printers operate at a high bit rate which may be greater than the maximum bit rate that the line is able to handle. In such cases the terminal must increase the bit rate before printing the data. When the output of a terminal appears on a VDU a high output bit rate is required so that the displayed information appears rapidly on screen. Many terminals incorporate a keyboard; this may be a

Table 1.2 ITU-T V recommendations

V1.	Equivalence between binary notation symbols and the significant conditions of a two-condition code.
V2.	Power levels for data transmission over telephone lines.
V3.	International alphabet No. 5 for data and message transmission.
V4.	General structure of signals for International Alphabet No. 5 for data and message transmission over public telephone networks.
V5.	Standardization of data signalling rates for synchronous data transmission in the PSTN.
V6.	Standardization of data signalling rates for synchronous data transmission on leased telephone-type circuits.
V10.	Electrical characteristics for unbalanced double-current interchange circuits for general use with IC equipment in data communications.
V11.	Electrical characteristics for balanced double-current interchange circuits for general use with IC equipment in data communications.
V13.	Answer-back unit simulators.
V15.	Use of acoustic coupling for data transmission.
V16.	Medical analogue data transmission modems.
V19.	Modems for parallel data transmission.
V20.	Parallel data transmission modems standardized for universal use in the PSTN.
V21.	200 baud modems standardized for use in the PSTN.
V22.	Standardization of data signalling rates for synchronous operation in the PSTN.
V22 bis.	Standardization of data signalling rates for synchronous data transmission on leased telephone circuits.
V23.	600/1200 baud modems standardized for use in the PSTN.
V24.	List of definitions for interchange circuits between data terminal equipment (DTE) and data circuit terminating equipment (DCE).
V25.	Automatic calling and/or answering on the PSTN, including disabling of echo suppressors on manually established calls.
V25 bis.	Protocols for the control of dialling procedures over the V24 interface using the normal data path.
V26.	2400 bits/s modem standardized for use in the PSTN.
V26 bis.	2400/1200 bits/s modem standardized for use in the PSTN.
V26 ter.	2400/1200 bits/s modem with echo cancellation standardized for use in the PSTN.
V27.	Modem for data signalling rates up to 4800 bits/s over leased circuits.
V27 bis.	4800 bits/s modem with automatic equalizer standardized for use on leased circuits.
V27 ter.	4800/2400 bits/s modem standardized for use in the PSTN.
V28.	Electrical characteristics for unbalanced double-current circuits.
V29.	9600 bits/s modem for use on leased circuits.
V30.	Parallel data transmission systems for universal use in the PSTN.

Table 1.2 (Cont'd)

V31.	Electrical characteristics for single-current interchange circuits controlled by contact closure.
V32.	9600 bits/s echo-cancelling modem for use on two-wire circuits.
V32 bis.	14 400 bits/s modem with echo cancellation using Trellis encoded modulation.
V32 ter.	19 200 bits/s modem. (Not ratified by the ITU-T.).
V33.	14 440 bits/s modem for use on four-wire circuits using Trellis encoded modulation.
V34.	28 000 bits/s modem.
V35.	Data transmission at 48 kilobits/s using 60 to 108 kHz group band circuits.
V36.	Modems for synchronous data transmission using 60 to 108 kHz group band circuits.
V37.	Synchronous data transmission at a data signalling rate higher than 72 kbits/s using 60 to 108 group band circuits.
V40.	Error indication with electromechanical equipment.
V41.	Code-independent error-control system (CRC 16-bit error check).
V42.	Standardization of modems that use non-synchronous-to-synchronous conversion techniques to check for errors.
V42 bis.	Specification for data compression used in conjunction with V42.
V50.	Standard limits for transmission quality of data transmission.
V51.	Organization of the maintenance of telephone-type circuits used for data transmission.
V52.	Characteristics of distortion and error-rate measuring equipment for data transmission.
V53.	Limits for the maintenance of telephone-type circuits used for data transmission.
V54.	Local and remote loopback testing.
V55.	Specification for an impulsive noise-measuring instrument for telephone-type circuits.
V56.	Comparative tests of modems for use over telephone-type circuits.
V57.	Comprehensive data test set for high data rate signalling rates.

QWERTY keyboard as used by typewriters, teleprinters and computer terminals, or it may be a keypad that has only a few selected functions/numbers available as, for example, on a bank/building society cash dispenser. The terminal must be able to convert the input data into the particular code that the distant host computer is able to recognize and so some form of code conversion is often necessary. Workstations are employed ever increasingly and these machines lie somewhere in between personal computers and minicomputers and offer high-resolution graphics capabilities.

Terminals send their data to line in one of two ways.

(*a*) The transmission of the data is under the control of a human operator. All unbuffered non-synchronous terminals are of this type. As the keyboard is operated the chosen characters are sent directly to line. This system requires that a point-to-point line is provided and the distant computer must have an input buffer (or store) and must *poll* each line regularly to detect any incoming data.

(*b*) The terminal includes a buffer store and input data goes into the store and is held there until the terminal is signalled by the distant computer that transmission may begin. This type of terminal is more expensive than the former but it allows the use of various line-sharing techniques (p. 92) and hence reduces the number of expensive lines and line equipments that are necessary.

2 Communication Techniques

The public switched telephone network (PSTN) was originally designed for the transmission of analogue speech signals. Signals are band-limited and trunks are routed over the core network. Multi-channel telephony systems are digital systems that employ a technique known as *pulse code modulation* (PCM). Eventually, the access network will also be digital but not for some years yet. In the meantime, however, many data circuits will continue to be routed via one, or more, analogue circuits. BT offer services, known as Kilostream and Megastream, which make high-speed digital circuits available to users but these can only be used for dedicated circuits between two particular points.

When a digital signal is transmitted over a telephone line the effect of the line attenuation and phase shift will be to distort the pulse waveform. This effect worsens with increase in both the bit rate and the length of the line and it will severely restrict the length of line that is usable at any particular transmission speed.

A digital signal transmitted over a telephone circuit contains significant energy at zero frequency and at low frequencies. Figure 2.1 shows the energy spectra for both speech and data signals, where τ is the time period of each bit. Clearly, most of the energy is concentrated at frequencies that are numerically lower than the bit rate of $1/\tau$ bits/s.

Unfortunately, digital signals cannot be transmitted over an analogue telephone circuit, whose bandwidth is 300 Hz to 3400 Hz, without considerable distortion occurring. This will make it difficult for the receiver to differentiate the received signal from noise. This distortion occurs because (*a*) most of the power content of the signal will not have been transmitted, and (*b*) a digital waveform contains some components at high frequencies and these are also lost. For transmission over all but fairly short distances the digital signal must first be converted into analogue, voice-frequency, form by the use

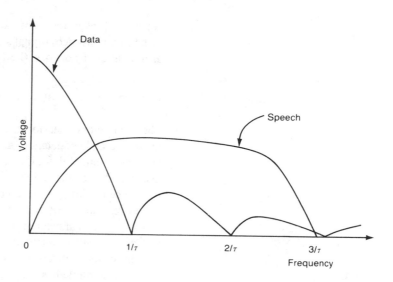

Fig. 2.1 Energy spectra of both speech and data signals.

of an equipment known as a modem. It is then known as a voice-frequency signal.

Transmission Lines

A transmission line consists of a pair of copper conductors that are separated from one another by a dielectric. There are two main types of transmission line in common use: they are the two-wire or twin-line shown in Fig. 2.2(a) and the coaxial line shown in Fig. 2.2(b). The two-wire line may be either open-wire, e.g. the final distribution from a telephone pole to a house, or it may be one pair of conductors in a multi-pair telephone cable. The telephone cables used for short-distance circuits are mainly of the star-quad type in sizes ranging from 14 pairs to 1040 pairs. The term star-quad means that the cable is made up by grouping the conductors in fours. Local line distribution cables are normally unit-twin cables; in this type of cable the conductors are twisted together in pairs and then a number of pairs (typically 50 or 100) are grouped together to form a unit. A number of units are then combined to form the cable.

The type of coaxial cable used in the home to connect the television receiver to its aerial has only one pair, but the coaxial cables used in the telephone network have two, or more, coaxial pairs grouped together, with star-quad cables placed in the gaps, or interstices, to form a complete coaxial cable.

Optical-fibre cables are commonly employed in the core telephone network. An optical fibre cable consists of a cylindrical glass core that is surrounded by a glass cladding and it is able to transmit a light wave with very little loss of energy. Optical-fibre cables offer the following advantages over copper transmission lines: (a) light-weight, small-dimensioned cables; (b) very wide bandwidth; (c) freedom from electromagnetic interference; (d) low attenuation; (e) high reliability and long life; (f) cheap raw materials; and (g) negligible crosstalk

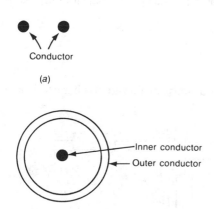

Fig. 2.2 (a) Two-wire line or twin-line and (b) coaxial line.

between fibres in the same cable. Optical fibre is particularly suited to the transmission of digital signals and it is often used for the cabling of a local area network (LAN).

Primary Coefficients of a Line

The conductors that form a pair in a telephone cable have both resistance and inductance, and they have capacitance and leakance between them. All four of the primary coefficients are uniformly distributed over the length of the line. The resistance R is the loop resistance in ohms of a one-kilometre length of the line, i.e. the sum of the resistances of each conductor. For a two-wire line the two conductor resistances are normally of the same value but the inner and outer conductors of a coaxial pair have different resistances because of their different cross-sectional areas. The inductance L of a line is also quoted as the total series inductance of both conductors, i.e. the loop inductance, and it is quoted in henrys per kilometre. The capacitance C is the total capacitance between a one-kilometre length of the two conductors and it is quoted in microfarads per kilometre. Lastly, the leakance G (often called the conductance) of a line represents the leakage of current between the two conductors. This leakage occurs partly because the insulation resistance between the conductors is not infinite, and partly because current must be supplied to supply the power losses in the dielectric as the line capacitance is charged and discharged.

A transmission line can be represented by a large number of the networks shown in Fig. 2.3. The line is considered to consist of a very large number of very short lengths dl of line connected in cascade. Each short section has a total shunt capacitance Cdl and total shunt leakage Gdl. For convenience, the total series resistance Rdl and the total series inductance Ldl are shown in the upper conductor only and are split into two equal halves.

Secondary Coefficients

The secondary coefficients of a transmission line are its characteristic

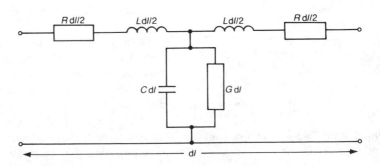

Fig. 2.3 Representation of a transmission line.

impedance, its attenuation coefficient, its phase-change coefficient, and its velocity of propagation.

Characteristic Impedance

The characteristic impedance Z_0 of a transmission line is the impedance that is determined by its physical dimensions and by the permittivity of the dielectric between the two conductors. The input impedance of a line will be equal to its characteristic impedance if the impedance of the load connected to its output terminals is also equal to the characteristic impedance. This is shown by Fig. 2.4. As for any other circuit, the input impedance is the ratio (voltage across input terminals)/(current flowing into terminals). Typical figures for the characteristic impedance of a line are (a) audio two-wire cable, 200 to 800 Ω, and (b) coaxial cable 50 to 75 Ω.

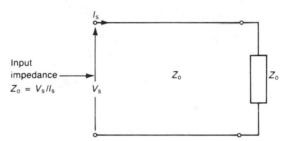

Fig. 2.4 Characteristic impedance of a line.

Attenuation

When a current, or a voltage, is propagated along a transmission line its amplitude is progressively reduced or attenuated. The percentage reduction in amplitude is exactly the same in each kilometre of the line. If, for example, the input voltage is 12 V and 10% is lost in the first kilometre the voltage entering the second kilometre of line will be 10.8 V. In the second kilometre length of line 10% of this voltage will be lost so that the voltage entering the third kilometre length will be 9.72 V, and so on. This means that the current, and voltage, waves decay in an exponential manner as they travel along a line, see Fig. 2.5.

The attenuation of a line is usually quoted in decibels per kilometre and Fig. 2.6 shows how, typically, the attenuations of (a) a star-quad cable, and (b) a coaxial cable vary with frequency. Clearly, the attenuation of a cable pair increases with increase in the frequency of the transmitted signal.

Phase-change Coefficient

A current, or voltage , wave travels along a line with a finite velocity and so the signal at the end of a line will have a phase lag with respect

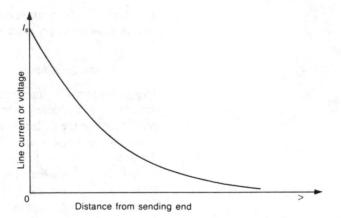

Fig. 2.5 Attenuation of a line.

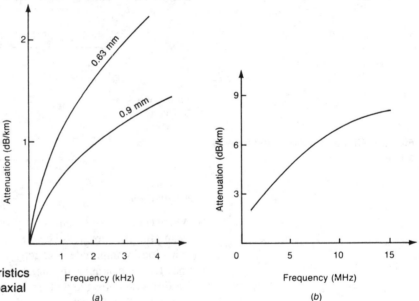

Fig. 2.6 Typical attenuation–frequency characteristics of (a) star-quad cable and (b) coaxial cable.

(a)

(b)

to the input current. The phase difference between the voltages 1 km apart is known as the phase-change coefficient β of the line. β is measured in radians per kilometre. In each kilometre length of the line there will be the same phase shift and hence for a line that is l kilometres long the total phase difference will be equal to βl rad. Figure 2.7 shows how the phase-change coefficient of a typical audio-frequency line varies with frequency.

Velocity of Propagation

The phase velocity v_p of a line is the velocity with which a sinusoidal wave travels along that line. The phase velocity is equal to the angular

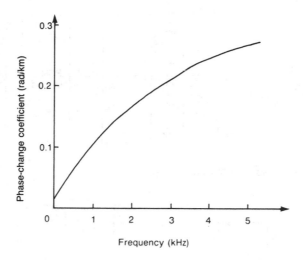

Fig. 2.7 Phase-change coefficient of a line.

velocity of the signal divided by the phase-change coefficient, i.e.

$$v_p = \omega/\beta \text{ m/s}. \tag{2.1}$$

A digital data waveform contains components at more than one frequency and each of these components will propagate along a line with its own phase velocity of propagation. For all the component frequencies of the digital waveform to travel with the same velocity and so arrive at the far end of the line at the same time, the phase-change coefficient β must be a linear function of frequency, i.e. the ratio ω/β must be constant at all frequencies. If it is not, and this is always the case at audio frequencies, the various frequency components will travel with different velocities and so take different times to travel over the line. The result is that the data waveform suffers *group-delay distortion*.

The *group velocity* v_g of the data signal is equal to

$$v_g = (\omega_2 - \omega_1)/(\beta_2 - \beta_1) \text{ km/s} \tag{2.2}$$

The *group delay* is the product of the length of the line and the reciprocal of the group velocity. Typically, the group delay of an audio-frequency cable is of the order of 7 μs/km.

Two-wire and Four-wire Presented Circuits

The terms two-wire and four-wire refer to the line circuit that is presented to the modem. A link can only be operated on a two-wire basis over its entire length if the line length is very short and/or the bit rate is very low. Higher transmission speeds require the use of modems at each end of the line and Fig. 2.8 shows a two-wire presented circuit. Each modem is connected by a two-wire line in the access network to the nearest PCM terminal from which it is transmitted, in the same way as speech signals, over the PCM system to the other PCM terminal.

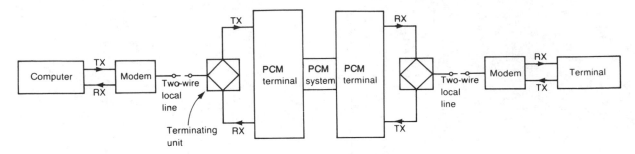

Fig. 2.8 Two-wire presented circuit.

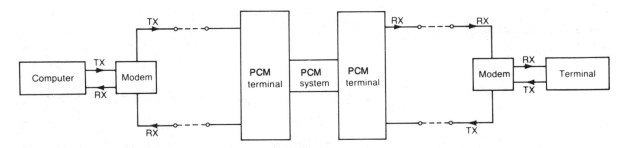

Fig. 2.9 Four-wire presented circuit.

When a data circuit is said to be four-wire presented it means that two pairs of conductors are extended from the PCM terminal right up to the modem as shown by Fig. 2.9. An echo in a telephone circuit is caused by the reflection of signals back to the transmitting end. Echoes normally have little effect upon speech but they are disruptive to full-duplex high-speed data circuits because the echoes appear as a form of corrupt incoming data. The main sources of echoes are the terminating units employed at each end of a telephone circuit to convert from two-wire to four-wire operation, and vice versa. Most troublesome are echoes from the far end terminating unit. Some of the more complex modems use echo-cancellation techniques to overcome this difficulty. This means that the transmitter feeds a copy of the outgoing line signal back into its own receiver; the receiver is then receiving signals from both the local, and the distant, transmitters simultaneously. The separately fed local copy is subtracted from the combined signal to leave only the wanted remote signal.

Digital Data Waveforms

A digital data signal consists of some combination of bits that represents, using a given data code, a number of characters. Each bit is of the same time duration and the number of bits transmitted per second is known as the *bit rate*. The data signal may be *unipolar* when binary 1 is represented by a voltage and binary 0 by zero volts, or it may be *bipolar* when binary 1 is represented by a voltage of one polarity and binary 0 by a voltage of the opposite polarity. In

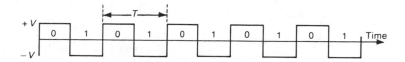

Fig. 2.10 Square data waveform.

most systems binary 0 is represented by a positive voltage, and binary 1 by a negative voltage. Both voltages may be anywhere in the range ± 3 V to ± 15 V.

When the data waveform consists of alternate 1s and 0s it has a square waveshape as shown by Fig. 2.10. The periodic time T of the waveform is the time interval between the leading edges of consecutive pulses and it is equal to the reciprocal of the bit rate. Any rectangular waveform contains components at a fundamental frequency and at a number of harmonics of that fundamental. In the case of the square waveform the harmonics are all odd, i.e. the third, the fifth, the seventh, etc., harmonics are all present. Clearly, in this case, the average value, or d.c. component, of the square wave is zero since there are equal numbers of positive and negative pulses, but other rectangular waveforms will have a non-zero d.c. component. In the periodic time of the square waveform two bits occur, 1 and 0, and hence the fundamental frequency is equal to $1/T$, and it is therefore numerically equal to one-half of the bit rate.

Fig. 2.11 Data waveform with alternate pairs of 1s and 0s.

If the data waveform consists of alternate pairs of 1s and 0s, as in Fig. 2.11, four bits occur in the periodic time T and so the fundamental frequency is now one-quarter of the bit rate. If the data waveform consists of consecutive 1s or 0s, as in Fig. 2.12, the fundamental frequency is zero hertz. The data characters transmitted over a circuit consist of many different combinations of 1s and 0s but the most rapidly changing waveform — and hence the one with the highest fundamental frequency — is the waveform with alternate 1 and 0 bits. Thus the fundamental frequency of a data waveform will vary from zero hertz to a maximum of one-half the bit rate. This means, of course, that the higher the bit rate the higher will be the maximum fundamental frequency and the greater will be the losses experienced in transmitting the signal over a line.

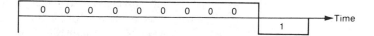

Fig. 2.12 Data waveform with consecutive 0s.

Example 2.1

Determine the maximum fundamental frequency of a 1200 bits/s data waveform. What other frequencies are also present?

Solution

The maximum fundamental frequency is 1200/2 = 600 Hz.   (*Ans.*)

The other frequencies that are present are:

(*a*) the third harmonic 1800 Hz;
(*b*) the fifth harmonic 3000 Hz;
(*c*) the seventh harmonic 4200 Hz, etc.   (*Ans.*)

Baseband Coding

For local short interfaces with, or without, a *line driver* and for a local area network (LAN) the d.c. data signal may be transmitted without the use of a modulated carrier wave. This is often known as baseband transmission. The bipolar data signal may be transmitted as it stands (this is known as non-return-to-zero, NRZ), or it may be coded before it is transmitted. The use of NRZ transmission can lead to various difficulties because:

(*a*) no synchronization clock is available because there may well be a lack of transitions if, for example, several consecutive 1 or 0 bits should occur;
(*b*) a d.c. path is essential to transmit the d.c. component; and
(*c*) low-frequency noise cannot be filtered out.

These problems can be overcome by the use of some kind of baseband coding. Whatever method of coding is employed its objectives are:

(*a*) to give zero residual d.c. component, i.e. the numbers of positive and negative states must balance out over some short period of time;
(*b*) to occupy the minimum possible bandwidth;
(*c*) to achieve a high transmission rate;
(*d*) to provide adequate timing information; and
(*e*) to be fairly easy to encode and to decode.

NRZ is the simplest code to implement and it is generally used by data terminals. The main encoding methods that are employed are Manchester, differential Manchester, WAL2 and Miller, and each of these codes will be discussed briefly.

The Manchester code carries checking information with the data stream which makes synchronization easier. The time interval for each bit is divided into two halves: the signal level in the first half represents the complement of the binary value of the data signal; the signal level in the second half represents the un-complemented binary value of the data signal. This means that there is always a transition from one state to another in the middle of each bit period and these transitions represent the data. A 1 bit is represented by a transition from positive voltage to negative voltage and a 0 bit is represented by a change from negative to positive voltage. If two bits of the same polarity are

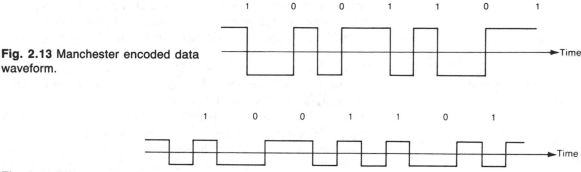

Fig. 2.13 Manchester encoded data waveform.

Fig. 2.14 Differential Manchester encoded data waveform.

next to one another a resetting transition is inserted between the two bits. Consider the ASCII character M or 1001101, its Manchester encoded waveform is shown in Fig. 2.13. The clock for the receiver can be extracted from the data stream because there is at least one transition in every bit period. Thus, the coded signal is both d.c. free and self-clocking, and since the absence of a transition indicates an error, it also provides inherent error detection. The maximum transition rate is twice the bit rate so that the required channel bandwidth is doubled. The system is employed for transmission over cables, both copper and fibre optic, and it is used in LANs.

The differential form of Manchester encoding avoids the need to know which state is which. In this system binary 0 is represented by the absence of a transistion between states in the middle of each bit period, and binary 1 is represented by the presence of such a transition. The state of the waveform is changed at the end of each bit period and this is shown by Fig. 2.14. Both of the Manchester codes are employed in LANs.

Another form of encoding that is well suited to transmission over two-wire pairs because its power spectrum is the inverse of the line's attenuation-frequency characteristic is known as WAL2. This encoding method has zero d.c. component but requires a bandwidth of about 2.5 times the uncoded signal. It is very similar to the Manchester code but it has the clock waveform shifted forward by 90 degrees.

The Miller code is used when bandwidth limitations are of prime importance since it reduces the maximum transition rate to be equal to the bit rate. Essentially, it is a modification of the differential Manchester code in which additional transitions are only included between consecutive 0 bits. Figure 2.15 shows the Miller code variation of the character M. The reduction in the bandwidth

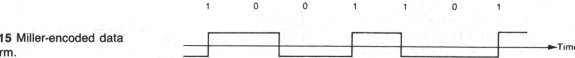

Fig. 2.15 Miller-encoded data waveform.

requirement is, unfortunately, offset by there being a d.c. component which can be quite large for some bit patterns. Also the code is harder to both encode and to decode than either of the Manchester codes.

The Effect of Lines on Data Signals

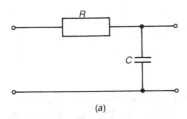

(a)

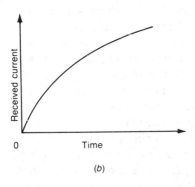

(b)

Fig. 2.16 (a) Representation of a line at low bit rates. (b) Variation with time of received current.

At low data rates over short distances the inductance and leakance of a transmission line have little effect upon a data signal and so the line may be represented approximately by the circuit shown in Fig. 2.16(a), where R is the total line resistance and C is the total line capacitance. Because of the need to charge the line capacitance the current at the receiving end of the line will not immediately rise to its final value when a pulse is applied to the input of the line. Instead, the received current will increase in an exponential manner as shown by Fig. 2.16(b). If the received current is not allowed sufficient time to reach its maximum value before the pulse ends the pulse waveform will not be received correctly. If the time taken for the received current to attain its final value is less than the time duration of a pulse the received pulse waveform will only be affected by the varying attenuation of the line at different frequencies. If, however, the *risetime* of the received current is greater than the bit length, distortion will take place. Figure 2.17 shows the effect on the waveform of the received current when the risetime is (b) less than, (c) greater than, and (d) much greater than, the bit duration.

The risetime is specifically the time taken for the received current to rise from 10% to 90% of its final, steady-state value. The risetime is equal to $2.2CR$, and the received current reaches its maximum value after a time equal to $4.5CR$ seconds.

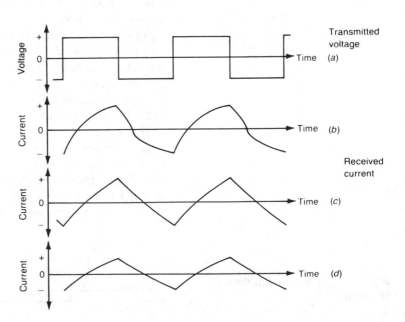

Fig. 2.17 Effect on the received current waveform when the risetime is (b) much less than, (c) greater than and (d) much greater than the bit duration.

Fig. 2.18 Effect of line attenuation on a data waveform.

When the data rate is high and the line length is long the line can no longer be regarded as a form of *RC* circuit. The attenuation of an audio-frequency cable increases with increase in frequency and so the various harmonics contained in the data waveform will be attenuated to a greater extent than is the fundamental. As a result the rectangular waveshape will be lost. The longer the length of the telephone line the greater will be the attenuation at each frequency with the result that the pulses become more and more rounded in shape as they travel down the line. This effect is shown by Fig. 2.18. The higher the bit rate the higher will be both the fundamental frequency and its harmonics and so the shorter will be the length of line at which satisfactory reception becomes impossible. However, direct d.c. data transmission is used for micro-computer systems and non-synchronous terminals over short distances at bit rates up to 9600 bits/s and at much higher bit rates in LANs.

The d.c. Component

A d.c. digital data signal cannot be transmitted over the PSTN because its d.c. component will be lost and this loss will distort the received signal waveform. If the bit rate is low the fundamental frequency of the data signal might also not be received nor, perhaps, will some of its harmonics. Removal of the d.c. component will always cause the data signal to shift either upwards or downwards, relative to the zero voltage axis, and this will make either the 1 or the 0 bits have a very small amplitude. The extreme case occurs when several successive 1 or 0 bits are transmitted and it is illustrated by Fig. 2.19.

There are three main reasons why the PSTN is unable to transmit the d.c. component of data signals. These are: (*a*) transmission bridges are used in telephone exchange equipment; (*b*) line-matching transformers are used to match different cables in the access network together; and (*c*) multi-channel PCM telephony systems are not able to transmit signals at or near zero hertz.

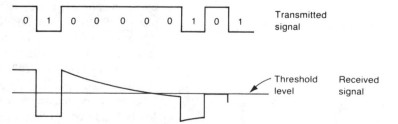

Fig. 2.19 Effect of the removal of the d.c. component of a data waveform.

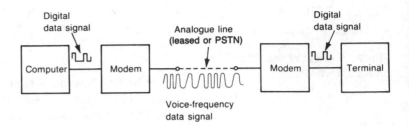

Fig. 2.20 The use of modems in a data circuit.

The Use of Modems in Data Circuits

When a data signal is to be transmitted over an analogue circuit, whether it is a dedicated leased circuit or a dial-up connection via the PSTN, the digital data signal must first be converted to analogue voice-frequency form. The required analogue-to-digital conversion is carried out in an equipment known as a *modem* at the transmitting end of the circuit. The v.f. signal is transmitted over the analogue line to a distant modem and here the signal is subjected to a digital-to-analogue conversion to restore it to its original digital form. The basic idea of the use of modems in an analogue data circuit is illustrated by Fig. 2.20.

Digital Modulation

The bandwidth of a commercial-quality speech circuit is 300–3400 Hz and this allows low-speed data links, operating at up to about 1200 bits/s, which have each bit represented by a single change in the modulated waveform to be transmitted. Higher bit rates require that the data stream is encoded before the modulation process is carried out. Adjacent bits may be paired together to form the *dibits* 00, 01, 11 and 10; each dibit can now be represented by a single change in the modulated waveform and hence such changes occur only *half* as often. This means that the baud speed on the line is only one-half of the bit rate. For an even greater reduction in the ratio (baud speed)/(bit rate) the bits making up the data stream can be grouped together in threes to form the *tribits* 000, 001, 010, etc., or into fours to form the quabits 0000, 0001, 0010, etc., or in fives to form the quinbits 00000, 00001, 00010, etc. Even 6-bit encoding (hexbits) is used for 14 400 bits/s Trellis-encoded systems.

The digital-modulation methods that are commonly used in data communication are frequency shift modulation (FSK), differential phase shift modulation (DPSK), and quadrature amplitude modulation (QAM). Table 2.1 gives the bit rates, and the relevant ITU-T recommendations for each of these types of modulation.

A modem that uses V42/V42 bis error correction can (in theory at least) quadruple the speed of V32 bis giving 57 600 bits/s. PCs however have a maximum output speed of 38 400 bits/s. V32 bis with MNP 5 data compression can give a bit rate of 19 200 bits/s.

Table 2.1

Modulation	ITU-T recommendation	Bit rates (bits/s)
FSK	V 21	300
FSK	V 23	1200/600
DPSK	V 22	1200/600/300
DPSK	V 26	2400
DPSK	V 26 bis	2400
DPSK	V 27	4800
DPSK	V 27 bis	4800/2400
DPSK	V 27 ter	4800/2400
QAM	V 22 bis	2400/1200
QAM	V 26 ter	2400/1200
QAM	V 29	9600/7200/4800
QAM	V 32	9600/4800/2400
QAM	V 32 bis	14 400
QAM	V 32 terbo	19 200
QAM	V 33	14 400/12 000
QAM	V 34	28 000

Frequency Shift Modulation

When a data waveform is used to frequency shift modulate a sinusoidal carrier wave the frequency of the carrier wave is switched between two values. The higher frequency is used to represent binary 0 while the lower frequency represents binary 1. The switching is carried out at the bit rate of the data signal. This is shown by Fig. 2.21. The frequencies that are used to represent binary 1 and binary 0 have been specified by the ITU-T and they are given in Table 2.2.

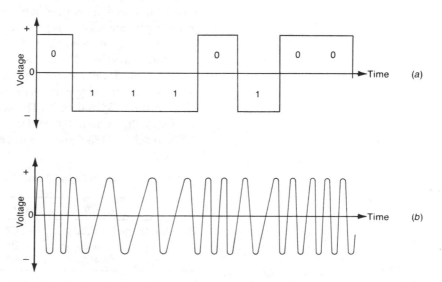

Fig. 2.21 Frequency shift modulation.

Table 2.2 ITU-T V21 and V23

Bit rate (bits/s)	Frequencies (Hz)	
	Binary 0	Binary 1
Up to 300	1180	980 } Different
	1850	1650 } directions
600	1700	1300
1200	2100	1300

The receiver has to be able to detect, at each sampling instant, which of the two frequencies is present. The higher the bit rate the further apart the two frequencies must be for the receiver to do this accurately and reliably. Because of this factor and the limited bandwidth made available on the PSTN the maximum bit rate for a FSK system is only 1200 bits/s.

This makes the system suitable for non-synchronous, but not for synchronous transmissions.

Differential Phase Shift Modulation

A phase modulation system operates by shifting the phase of a sinusoidal carrier wave between two different values to represent the digital data signal. The phase of the carrier is shifted by 180° each time the leading edge of a 0 bit occurs, but the phase of the carrier is not altered by a 1 bit. This basic form of phase-shift modulation is rarely used in practice since it is difficult to detect the received signal and convert it back into digital form. Instead, a variation, known as *differential phase-shift modulation* (DPSK) is employed. With this form of modulation *changes* in phase, rather than absolute values of phase angle, are used to represent dibits or tribits. Table 2.3 gives the ITU-T recommendations for the V22, V26, and V26 bis synchronous systems. These phase changes lead to eight carrier phases being generated, i.e. for the +45° changes used to represent the dibit 00 in the V26 bis system the phase angles are 45°, 90°, 135°, 180°, 225°, 270°, 315° and 360°. The other V26 bis phase changes lead to exactly the same phase angles.

The resulting DPSK signal is often represented by a constellation diagram ;

Table 2.3 ITU-T V22, V26 and V26bis

	Dibit	00	01	11	10
Phase change	V22	+90°	+0°	+270°	+180°
	V26	+0°	+90°	+180°	+270°
	V26bis	+45°	+135°	+225°	+315°

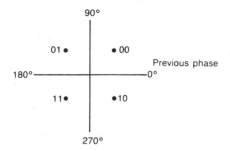

Fig. 2.22 Constellation diagram of a V26 bis signal.

this is a diagram that shows the amplitude and phase relationships that are associated with each combination of dibits, tribits, or quabits. Each point in the constellation diagram represents multiple bits of data. The higher the transmission speed of the signal the more complex will be the constellation. The constellation of a V26 bis signal is shown by Fig. 2.22. An 1800 Hz carrier is used to position the modulated wave in the middle of the frequency band where group-delay distortion is least. For random data the signal energy is uniformly spread over the band 600 Hz to 3000 Hz but always any frequency component that is below the carrier frequency has a corresponding frequency component above the carrier frequency; their frequency separation is numerically equal to the baud speed.

The bandwidth required to transmit a DPSK signal is given by equation (2.3), i.e.

$$\text{Bandwidth} = (\text{bit rate})/\log_2 n = (\text{bit rate})/(3.32\log_{10}n), \quad (2.3)$$

where n is the number of phase changes used.

Example 2.2

Calculate (*a*) the line baud speed, and (*b*) the necessary bandwidth, for a 2400 bits/s DPSK system.

Solution

    (a) Baud speed = 2400/2 = 1200 baud.   (*Ans.*)

    (b) Bandwidth = $2400/(\log_2 4)$ = 1200 Hz.   (*Ans.*)

The carrier frequency employed is 1800 Hz and so the occupied bandwidth is 1200 Hz to 2400 Hz.

A DPSK modem operating at 4800 bits/s encodes the data stream into tribits. Eight different combinations of tribits are possible and so eight phase changes are necessary. The ITU-T V27 values are given in Table 2.4. A 4800 bits/s data signal can now be transmitted over a line at a baud speed of 1600 baud in the bandwidth 1000 Hz to 2600 Hz. The constellation diagram of a V27 DPSK signal is shown by Fig. 2.23.

Table 2.4 ITU-T V27

Tribit	000	001	011	010	110	111	101	100
Phase changes	+45°	+0°	+135°	+90°	+225°	+180°	+315°	+270°

Quadrature Amplitude Modulation

Quadrature amplitude modulation (QAM) is a mixture of amplitude modulation and phase modulation that is employed at bit rates of 1200

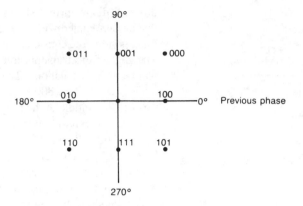

Fig. 2.23 Constellation diagram of a V27 signal.

bits/s and above. The data bit stream is divided into groups of four bits, known as quabits, and hence there are 2^4 or 16 different combinations: 0000, 0001, 0010, etc. A carrier comprises two signals at the same frequency but in phase quadrature with one another, i.e. mutually 90° out of phase. Hence, as one of the signals approaches its maximum value the other signal is approaching zero voltage. A number 2^m of phase changes are used and for each phase change there are a number 2^n of possible amplitudes.

In the V22 bis and V22 ter systems two bits change the amplitude, and the other two bits change the phase of the carrier. This gives 2^2 or 4 possible amplitudes which are, relatively, -1, $-1/3$, $+1/3$, and $+1$; and 2^2 or 4 possible phases and therefore a total of 16 different states. The 16-point constellation is shown by Fig. 2.24. At the receiver these 16 line states must be recognized correctly so that the original data stream can be reconstituted correctly. V22 uses carrier

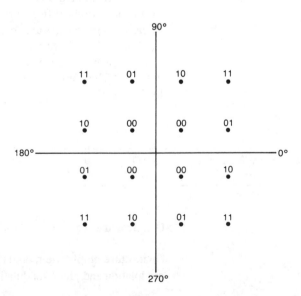

Fig. 2.24 Constellation diagram of a V22 bis/ter signal.

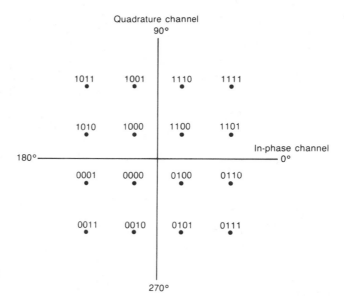

Fig. 2.25 Constellation diagram of a V29 signal.

frequencies of 1200 Hz and 2400 Hz to give full-duplex operation over a dial-up PSTN circuit or a two-wire leased circuit.

The V29 system uses one bit to vary the amplitude of the signal and the other three bits to vary its phase. Hence, there are two possible amplitudes and 2^3 or 8 possible phases which again gives 16 different states. The constellation diagram is shown by Fig. 2.25.

To achieve reliable results at higher bit rates such as 9600 bits/s, 14400 bits/s and 19800 bits/s over ordinary telephone circuits *trellis encoding* (TCM) is used. An encoding bit is added to each symbol to create specific bit stream patterns which the receiver is able to recognize. This means that 32 different five-bit symbols are transmitted to line. The receiver matches the received signals with the 32-point trellis-coded pattern and any difference found is an indication that an error has occurred. The practical effect is to give an increase, of about 3 dB, in the signal-to-noise ratio of the circuit. Once again the carrier frequency is 1800 Hz and the line baud rate is 2400 bauds. Figure 2.26 shows the constellation of a V32 9600 bits/s trellis-coded signal.

As the number of amplitude and phase change levels is increased both the transmitting and the receiving circuitry become more complex and hence the modem becomes more costly.

The Effects of Lines on Modulated Data Signals

When modulated data signals are transmitted over an analogue telephone line the combined effects of attenuation and group-delay distortion may distort the received signal to such an extent that the

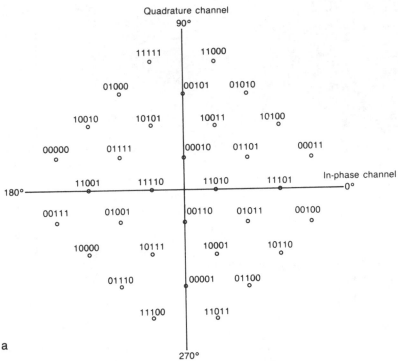

Fig. 2.26 Constellation diagram of a V32 trellis encoded signal.

receiver may not be able to reliably interpret the incoming data stream and then the bit error rate will increase.

For FSK signals the higher frequency will be attenuated more than the lower frequency and this may make it difficult for the receiver to determine when a 0 bit has been received. The effect is worse for 1200 bits/s systems than for 600 bits/s systems because of the higher 0 frequency employed. The effect of group-delay distortion is to delay the higher frequency more than the lower frequency and this lengthens the 1 bits and shortens the 0 bits. This is, of course, a form of bias distortion. In extreme cases the receiver may not be able to detect a 0 bit that is in between two 1 bits. With a DPSK system bias distortion is eliminated since the detection of the received signal is based upon the change in phase between the previous symbol and the present symbol. Since only relative changes need be detected the effects of group-delay distortion are reduced. QAM is also fairly tolerant of the effects of line attenuation and distortion since all high-speed modems incorporate adaptive equalizers to counteract these effects.

Transmission Media

Data circuits can be set up, on either a permanent or a temporary basis, via a number of different media. A link may be established by dialling on the PSTN, or on the PSDN, or by using a private circuit that has been leased from BT, Mercury, an international carrier, or some other provider. Except for very local connections, e.g. computer–peripheral, or LAN circuits, circuits are routed over PCM multichannel telephony systems. The telephone network was originally designed for voice communications and not for data communications which came along very much later. The bandwidth of the PSTN is limited and its transmission characteristics may not be as good as desired, (because of the access network's limitations), but the main drawback is that the call set-up time is too long for real-time applications. The bank cash dispensers, for example, require the overall response time to be no more than about 3 s, which is unattainable using the PSTN. Another disadvantage is that the cost of a PSTN call is based upon time duration as well as distance whereas it is generally considered that data tariffs ought to be based upon the volume of data transmitted.

BT and other carriers will provide leased lines which have been conditioned to improve their transmission quality. The term 'conditioned' means that the line has been *equalized* so that it has the same overall attenuation and/or group delay at all frequencies. For normal lines operating at low bit rates the equalization is carried out when the circuit is first set up. For high bit rates, at or near the limits of the line, it is necessary continually to monitor the line and to make real-time adjustments to the equalization. High-speed modems always incorporate an *adaptive equalizer*.

Circuits leased from BT are known as the A-line Services. Lines which are to be used for speech transmission are known as Speechline circuits, and lines that are to be used for data transmission are called Keyline circuits.

The public switched data network (PSDN), called Switchstream by BT, provides dial-up connections to other customers to the network. Also available are two point-to-point digital systems, known respectively as Kilostream and Megastream, which can provide wideband data circuits. Kilostream occupies one channel of the standard ITU-T 30-channel PCM system and provides a 64 kbit/s data circuit. The customer can use this circuit to provide one 64 kbit/s channel or, by the use of time-division multiplex (TDM) techniques obtain a number of lower-bit rate channels. Megastream occupies a complete PCM line system without the channel multiplexing terminal equipment, and it provides the customer with a high-speed 2.048 Mbit/s channel. This channel can be used by the customer as a single high-speed circuit or it can be multiplexed to give a large number of lower-speed channels. The specialized digital data circuits provide a much lower bit error rate than the analogue networks can offer, although this is often spoilt somewhat by the less-than-perfect local lines that may be used to connect the digital network to the customer's equipment.

When a network that has been designed for the transmission of digital signals is employed to transmit data signals it is not necessary to use modems. Instead, the effects of line attenuation and distortion

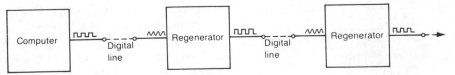

Fig. 2.27 Use of pulse regenerators in a digital line.

are overcome by the use of *pulse regenerators*. For the bit error rate to remain above the wanted minimum figure the signal-to-noise ratio on the line must be kept above a minimum threshold value. The digital data signal is applied to a pulse regenerator before the signal-to-noise ratio has fallen to too low a figure and the regenerator generates a new signal. The idea is shown by Fig. 2.27.

The use of a digital circuit for the transmission of data has a number of advantages: (*a*) less noise and distortion, (*b*) lower bit error rates, (*c*) no expensive modems are required, (*d*) circuit functions, such as multiplexing and switching, are simpler than their analogue equivalents, (*e*) the electronic circuits used in digital systems are all in IC form and are therefore generally cheaper and more reliable, and (*f*) other forms of signal, such as speech and television, can also be transmitted digitally as an integral whole with the data signal.

Kilostream

Kilostream is a BT service that provides a direct digital link between two, or more, points and it can provide full-duplex operation at 2.4, 4.8, 9.6, 48, or 64 kbits/s. BT provides an X21, V24 or V35 interface to link the customer's equipment to a network terminating unit (NTU). The NTU transmits using WAL2 encoding at either 12.8 kbit/s or 64 kbit/s over a four-wire line, to the nearest telephone exchange where the Kilostream terminal equipment is installed. Each Kilostream circuit offers a 64 kbit/s circuit in the manner shown by Fig. 2.28. The channel modulating equipment that the PCM system employs to convert speech signals into digital form is not provided; the digital data signals are applied straight to the encoding and combining circuitry. Here the data signals are changed into the PCM format and then 30 channels are combined, using time-division multiplex, to form a 2.048 Mbit/s digital signal. Synchronization signals are then added before the signal is transmitted over the bearer circuit.

Kilostream offers a cheaper alternative than leased lines using modems and it is increasingly popular wherever it is available and the customer has a requirement for a 64 kbit/s channel.

Megastream

Megastream is the name given by BT to a direct digital service that it provides between two points. The service operates at 2.048, 8, 34

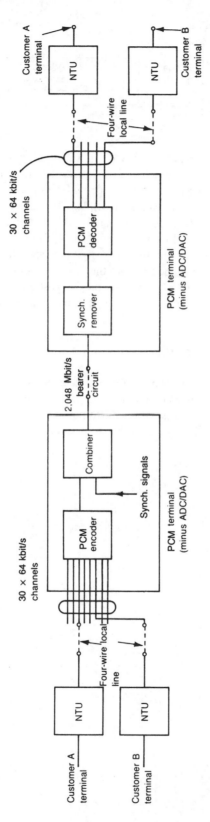

Fig. 2.28 Use of Kilostream to obtain 64 kbit/s circuits.

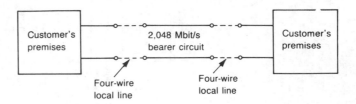

Fig. 2.29 Use of a Megastream circuit.

or 140 Mbit/s with an X21, V24 or V35 interface. Megastream utilizes a complete PCM bearer circuit minus the terminal multiplexing equipment and it provides the user with a high-speed circuit. This circuit can be multiplexed by the user to provide a large number of lower-speed channels. All of the capacity of the link need not be employed to carry data — if required some of the channels can also be used to carry digitized speech signals. A four-wire circuit is provided to link the customer's equipment to the nearest telephone exchange where the Megastream terminal equipment is situated. Figure 2.29 shows the basic concept of a Megastream circuit.

Satstream

Satstream is a digital data service offered by BT which uses capacity on a communication satellite to provide links to points within Western Europe. Each customer is provided with a small dish aerial to give him a radio link to a satellite.

3 Modems

The bandwidth of a commercial-quality telephone circuit is limited to 300–3400 Hz and this 3.1 kHz bandwidth is not wide enough to transmit digital data signals without the introduction of considerable distortion. It is therefore necessary to convert the digital data signal into the corresponding voice-frequency analogue signal before it is transmitted over the telephone network. At its destination the signal must be reconverted back into its original digital format. The necessary digital-to-analogue, and analogue-to-digital conversions are carried out by an equipment that is known as a *modem*. One modem is required at each end of a circuit. The ITU-T refer to a modem as a data communication equipment (DCE) and the EIA refer to it as a data circuit terminating equipment (also DCE). A modem may also be required to establish, maintain and terminate each connection that is set up via the telephone network, using either a leased circuit or a dialled-up connection via the PSTN. Automatic dialling and answering facilities are commonplace and some modems can also automatically restore communications after a link failure has occurred by finding an alternative channel. Modems operating at speeds in excess of 2400 bits/s also include error checking and correcting circuitry. The ITU-T recommendation V25 defines automatic calling procedures over a separate parallel interface and the V25 bis recommendation defines the protocols that must be used for control of dialling procedures over the V24 interface using the normal data path.

A modem uses some form of digital modulation to convert a digital data signal into voice-frequency form and the vast majority of modems comply with the relevant ITU-T recommendations. These are given by Table 3.1.

A fast modem allows files to be transferred in less time. V32 bis is 50% faster than V32 and 15 times faster than V26. The maximum speed of many public on-line databases and Electronic Mail services is usually 9600 bits/s, and sometimes only 4800 bits/s. There is then

Table 3.1

ITU-T recommendation	Bit rate (bits/s)		Type of modulation
	normal	fall-back	
V21	300	—	FSK
V22	1200	600	DPSK
	300	—	DPSK
V22 bis	2400	1200	QAM
V23	1200	600	FSK
V26	2400	—	DPSK
V26 bis	2400	1200	DPSK
V26 ter	2400	1200	QAM
V27	4800	—	DPSK
V27 bis	4800	2400	DPSK
V27 ter	4800	2400	DPSK
V29	9600	7200/4800	QAM
V32	9600	4800/2400	QAM
V32 bis	14 400	12 000/9600	QAM
V32 terbo	19 200	16 800/9600	QAM
V33	14 400	12000	QAM
V34	28 000	24 000/19 200	QAM

no need for a higher-speed modem to be employed. However, when two PCs are communicating with one another, the faster the bit rate employed the smaller will be the telephone bill. Most modems comply with several standards, particularly the faster modems which usually have downward compatibility with slower standards.

The reason for the use of several different modulation methods is that each of the methods is more suitable than the others, in terms of performance and cost, at different bit rates. It might appear that a baud speed of 3100 bauds could be employed but, in practice, the maximum line baud speed is 2400 bauds in order to minimize the group delay distortion that would otherwise occur at the lowest and highest frequencies in the available line spectrum. The carrier frequency is selected to position the modulated signal in the middle of the available bandwidth and it is usually in the region of 1700 Hz.

The management of a data network may require the modems to have a secondary channel as well as the data-carrying primary channel. While data is being transmitted over the primary channel the secondary channel can be used by the network management computer to monitor the system, to transmit commands, etc. Whereas the primary channel works at one of the speeds given in Table 3.1 the secondary channel operates at a much lower speed, typically between 75 bit/s and 150 bit/s. Some modems are said to be *intelligent*. This means that they incorporate a microprocessor which allows the modem to be used to monitor its own status, e.g. its power supply voltage(s), any faults that might occur, and the analogue signal's signal-to-noise ratio, and report it back to a network management control located elsewhere.

A modem is connected to its computer or terminal by means of an *interface*. This consists of plugs, sockets, pins and a cable that must be both electrically and mechanically compatible with each other and the equipments that are to be interconnected. Electrical compatibility means that both equipments must represent the binary states 1 and 0 by the same voltages; very often these voltages are either ± 6 V, or 0 V and $+5$ V.

The ITU-T recommendation V24 defines an interface for the connection of a computer or a terminal to a modem for serial transmission. The use of the V24 interface, or its US equivalent the EIA (Electronic Industries Association) specification EIA 232E, ensures that a system designer can be confident that two equipments bought from different manufacturers can be connected together and will then be able to exchange data. The V24 recommendation concerns itself only with electrical signals and it does not define any mechanical standards, but the EIA 232E specification defines both. V24 is normally used in conjunction with the ISO (International Standards Organization) D-type plug, the most common version of which has 25 pins. The electrical characteristics specified by V24 define only the set of control signals and the manner in which they ought to be used. Some modems are provided with an automatic dial back-up facility in which the modem continuously monitors the leased line. If the line should fail the modem will automatically place a dialled call over the PSTN to its opposite number. This other modem will answer the call automatically, check the identity of the calling modem, and then complete the dial-up connection to re-establish the connection. The calling modem will continue to monitor the leased line and immediately it is restored to service will switch the connection back to it.

Many modems incorporate built-in diagnostic testing facilities which follow the ITU-T V54 recommendations. These test facilities include the following.

(a) *Self-test*. This method, shown in Fig. 3.1(*a*), tests the modem hardware. The analogue terminals are looped together and when the modem is first switched on it automatically generates a predetermined bit stream which it compares with the data received after passing through the analogue loop. If no faults are found the modem goes into its operational state. Self-testing has the advantage that it is not necessary for an operator to have to type characters and then check the resulting display on a VDU.

(b) *Local analogue loopback*. The analogue terminals of the modem are connected together so that all the data originating from the terminal is first modulated, and then demodulated, by the modem before it is returned to the terminal. The method, shown in Fig. 3.1(*b*), allows the correct operation of both the modem and the terminal—modem interface to be tested.

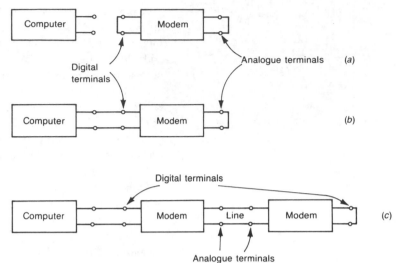

Fig. 3.1 Modem testing: (a) self-test, (b) local analogue loopback and (c) remote digital loopback.

(c) *Remote digital loopback.* The distant modem has its digital terminals looped together as shown in Fig. 3.1(c) so that all the data received by it is transmitted back to the originating terminal. The method checks the correct operation of both modems and also of the telephone line.

The choice of a modem for use on a particular data circuit is determined by both the required bit rate and economics. The higher the bit rate required the more expensive will be the modem. It is, of course, necessary to ensure that the modems at each end of a circuit operate to the same ITU-T recommendations. Operation at the same bit rate does not necessarily ensure this, as a reference to Table 3.1 will show. If half-duplex operation is required the modem employed must be capable of *turnaround* operation. Generally, this means that a high-speed modem is required. The complexity, and hence high cost, of a high-speed modem arises from the need to (a) compress the signal into the relatively narrow-band analogue channel, (b) minimize the effects of line noise, distortion and echo, and (c) avoid causing interference to other signals in adjacent pairs in the telephone cable. Another factor to be considered is the distance between the two modems. If the length of the line is short, say up to about 25 km, and it has a continuous d.c. path, a *short-haul modem*, a *line driver* or a *modem eliminator* can be used. There are no ITU-T recommendations for such devices but they are considerably cheaper than the higher-speed modems that have been designed to operate over long-distance circuits.

Modems operating at high bit rates, such as 9.6 kbit/s or more, may have more than one digital input port. Such modems have integral time-division multiplexers which allow the data input to the different digital ports to be combined before transmission as a voice-frequency signal over the telephone line.

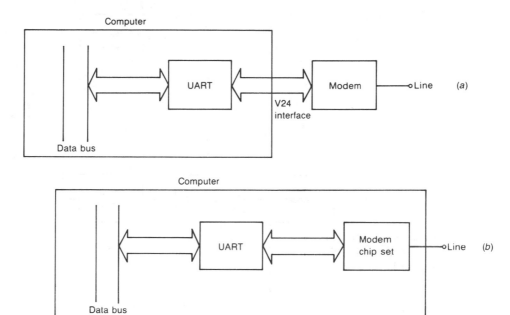

Fig. 3.2 (*a*) Use of a UART and a modem. (*b*) Internal modem saves a power supply and an interface.

High-speed modems are generally downward compatible, e.g. a 4.8 kbit/s modem can also work at 2.4 kbit/s and 1.2 kbit/s. Usually, the calling modem will recognize the maximum speed of the called modem during the handshake procedure. If this speed is less than the maximum speed of the calling modem, the calling modem will automatically reduce its speed to that of the called modem. This is known as 'fallback'. If, for example, a poor quality circuit will not permit two modems to work at 9600 bits/s they may negotiate a fall-back to 4800 bits/s or even 2400 bits/s. Some modems may remain at the fall-back speed if the line quality should improve but others, which incorporate a protocol known as MNP10, are able to adjust their speed to the highest bit rate that will give an acceptable BER. Such intelligent operation is nearly always associated with auto-dialling and auto-answering facilities.

Increasingly, semiconductor manufacturers are bringing onto the market single-, two-, or three-chip modems which need only the addition of a few external components. This trend, which is most marked at the lower bit rates, allows physically smaller, cheaper and more reliable modems to be designed. This has allowed some modems to be produced as an integral part of a terminal with the consequent advantages of eliminating the need for (*a*) a separate modem power supply and (*b*) a V24/EIA 232E interface, as shown by Figs 3.2(*a*) and (*b*). The block marked as UART represents a universal asynchronous receiver/transmitter chip and its purpose is to convert the parallel data from the computer bus into serial form.

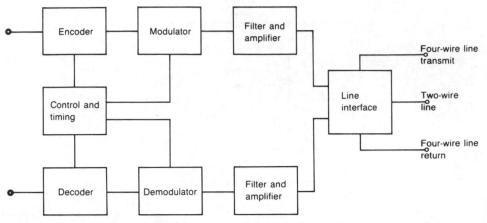

Fig. 3.3 Basic block diagram of a modem. Note: neither the encoder nor the decoder is needed for an FSK modem.

Types of Modem

The basic block diagram of a modem is shown in Fig. 3.3. The digital data input from the computer or terminal is applied to an encoder (unless it is a FSK modem), where the bit stream is encoded into dibits, tribits or quabits. The encoded signal is then applied to the modulator where a digitally modulated waveform is produced. The modulator output will contain various unwanted frequency components and so it is band limited by a filter and then amplified before it is passed on to the line interface. At this point the v.f. signal is applied to the line, which is either two-, or four-wire presented. In the receiving direction the incoming v.f. signal is filtered and amplified to remove all unwanted signals and noise before it is passed to the demodulator. The signal is demodulated to produce an encoded digital output and this is applied to the decoder whose function is to restore the received data into the wanted single-bit format.

V21 Modem

A V21 modem operates at 300, 600 or 1200 bit/s using FSK. Operation at 300 bit/s is full-duplex over the PSTN but the higher speeds of 600 and 1200 bit/s can only support half-duplex operation. Several ICs are now available that can perform most, if not all, of the functions of such a modem. It is only necessary to add a few external components, such as resistors, capacitors and a piezo-electric crystal to set the clock frequency. V21 is the only standard in which the bit rate is the same as the baud speed.

V22 Modem

A V22 modem operates either non-synchronously or synchronously at 600/1200 bit/s over either the PSTN or a leased line and it can give both half-duplex and full-duplex working. Each channel has a line baud speed of 600 bauds. There are three possible variants to the V22 recommendation but only two of them are commonly employed. These are: (*a*) Variant A, 1200 bit/s plus 600 bit/s, both synchronous; and (*b*) Variant B, as Variant A plus 1200/600 bit/s non-synchronous.

The lower channel operates with a carrier frequency of 1200 Hz and the upper channel with a carrier frequency of 2400 Hz. When signals are transmitted in the upper channel a 1800 Hz guard tone is also transmitted to disable any signalling unit that might have been on line when the core network employed analogue techniques. The modulation method employed is DPSK with the dibits 00, 01, 11 and 10 being represented by phase changes of $+90°$, $0°$, $+270°$ and $+180°$, respectively. A baud speed of only 600 bauds is needed to transmit a bit rate of 1200 bits/s.

V22 bis Modem

A V22 bis modem is able to work full-duplex over both the PSTN and a leased circuit at a bit rate of either 2400 bit/s or 1200 bit/s; the modulation method is either 16-phase QAM or four-phase QAM, respectively. In both cases the line speed is 600 baud. Full-duplex operation is made possible by the use of *frequency-division multiplex* (FDM). The usable telephone line bandwidth is divided into upper and lower frequency bands using carrier frequencies of 2400 Hz and 1200 Hz. The V22 bis recommendation specifies a handshake at the start of an incoming call that allows a receiving modem to switch automatically to the required speed of either 2400 bit/s or 1200 bit/s. The modem incorporates an adaptive equalizer.

V23 Modem

V23 specifies one of the earliest modems to be employed. Using FSK at a bit rate of either 600 bit/s or 1200 bit/s half-duplex operation is possible over a PSTN dial-up connection and full-duplex operation over a four-wire presented leased line. It provides 75 bits/s in the reverse direction of transmission and it was introduced for use in conjunction with viewdata systems such as BT's Prestel.

V26 Modem

The V26 recommendation is for 2400 bit/s operation using DPSK. The original recommendation allows only full-duplex operation over a leased line but the later variants, i.e. V26 bis and V26 ter, provide a 1200 bit/s fall-back facility with V26 bis providing half-duplex operation, and V26 ter providing full-duplex operation, over the PSTN.

V27 Modem

A V27 modem can operate at a bit rate of either 4800 bit/s or 2400 bit/s over either the PSTN or a leased line. V27 gives just full-duplex operation over a leased circuit, V27 bis half-duplex operation over the PSTN and full-duplex operation over a leased line, while V27 ter gives just half-duplex working over the PSTN.

Eight-phase DPSK is employed with a line baud speed of either 1600 baud, or 1200 baud. The carrier frequency is 1800 Hz and the occupied bandwidth is either 1000 to 2600 Hz or 1200 to 2400 Hz.

V29 Modem

A V29 modem operates at bit rates of 9.6, 7.2 and 4.8 kbit/s providing half-duplex operation over two-wire presented, and full-duplex operation over four-wire presented lines using either the PSTN or a leased circuit. The modulation method that is employed is QAM with a line baud speed of 2400 bauds and a carrier frequency of 1700 Hz. The occupied bandwidth is 500 to 2900 Hz.

At these high bit rates signal distortion is a serious problem due to noise, echoes, jitter of the modulated carrier, and transient phase-gain hits. The adverse effect of this distortion is to blur the points in the constellation diagram and so make it more difficult for the receiver to interpret, and hence reproduce the incoming data correctly. An *adaptive equalizer* is an essential part of this kind of modem. The use of trellis-coded QAM (TCM) allows 9.6 kbit/s full-duplex operation over the PSTN and 19.2 kbit/s operation over conditioned leased circuits. The correct reception of 9.6 kbit/s data over a 3 kHz channel involves the receiving modem making decisions between the 16 possible phase and amplitude states of the received carrier at a symbol rate of 2400 bauds. The 16-state constellation diagram is shown in Fig. 3.4, (see also Fig. 2.25). The ringed sub-set of four states is used for bit rates of 4800 bit/s giving the advantage that the average transmitted power is the same at both bit rates.

The V29 recommendation also includes an optional TDM multiplexer which allows several terminals to share the same leased line. The 9.6 kbits/s bearer circuit can be divided into two 4.8 kbit/s, or four 2.4 kbit/s channels, or various other combinations thereof. If the terminal attached to one of the digital channels is non-active the modem still assigns a time slot to it and then sends filler signals instead of data. The modem interleaves the bits from each terminal to produce a multiplexed bit stream that then modulates the carrier.

V32 Modem

A V32 modem operates at a bit rate of 9.6 kbit/s with a fall-back speed of either 4.8 kbit/s or 2.4 kbit/s using QAM to give full-duplex

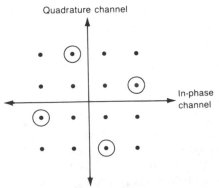

Fig. 3.4 Constellation diagram of a V29 modem.

operation at 2400 bauds over a dial-up connection via the PSTN. Trellis coding is employed at 9600 bits/s, this, similar to basic QAM, works by translating data constellation patterns but rather than looking at just one point at a time it compares successive points to determine the correct location of each point. Trellis encoding is performed in the transmitting modem and trellis decoding in the receiving modem. The decoding process is achieved by comparing the distance of each received signal point with the constellation template. This involves the use of about 1000 computer processing steps for each point and it is the reason why modems that operate at even higher bit rates, such as 14.4, 16.8, and 19.2 kbit/s are very expensive. Trellis encoding is not used at the fall-back speeds of 4800 and 2400 bits/s.

The V32 recommendation states that the modem must incorporate a circuit known as an *echo canceller*. The modem does not use frequency-division techniques to achieve full-duplex operation over a two-wire presented circuit as, because of the large amount of data to be sent, the entire 3.1 kHz bandwidth of a telephone line is needed in both directions of transmission. This means that a V32 modem must be able to handle superimposed signals and have the ability to extract the weaker received signal from the stronger transmitted signal. The problems involved in doing this are made more difficult by the presence of echoes that are generated in the transmission path. The detection, and then the subtraction, of near-end echoes can be achieved in a V32 modem by their internal signal processors. Far-end echoes, however, present a more difficult problem and only the latest, and most expensive, modems have the capability to deal with them. An adaptive equalizer is also included in the modem circuitry.

A V32 modem has an advantage over other types of modem with regard to its dial back-up feature. Modems of V29 or lower standard must disconnect the line currently used if and when contact with the distant terminal is lost and then immediately dial the same terminal via the PSTN. This is not the case with a V32 modem; if the leased line should fail the modem will automatically reconnect the link to maintain continuous service. The use of V32 modems enables the users to take full advantage of the flexibility of the PSTN to provide instant high-speed circuits between any two desired locations. Some V32 modems are employed to monitor a 64 kbit/s digital data link. If such a link should fail the modem will automatically dial a stored telephone number and then continue service on the dialled back-up line. Once the digital link is restored to service the modem will switch the link back to the digital line and drop the PSTN connection. Most V32 modems are provided with V42 error correction and V42 bis data compression (page 157).

V32 bis Modem

A V32 bis modem can operate at 14 400 bits/s with a line baud speed of 2400 bauds, while permitting the use of lower speeds such as

12 kbits/s, 9.6 kbits/s, 7.2 kbits/s and 4800 bits/s if the circuit's quality should be poor. The additional speeds are made available by the generation of different combinations of amplitudes and phases during the modulation of the carrier.

At 14 400 bits/s there are 128 different combinations of amplitudes and phases for each baud. This number is sufficient to encode seven data bits in each baud. Because there are so many amplitudes and phases the differences between them are small and hence errors are likely to occur. This means that error detection and correction is essential. When V42 bis data compression is employed in conjunction with V32 bis a maximum data rate of 57 600 bits/s is theoretically possible, but line noise and distortion will reduce this figure somewhat. Because most of the available bandwidth is used, full duplex operation is obtained by the use of echo-cancellation techniques; these make it possible to transmit data simultaneously in both directions in the same bandwidth.

V32 terbo Modem

A V32 terbo modem is able to work at 19.2 kbits/s using an extension of the V32 constellation; both convolutional encoding and Viterbi decoding are employed to reduce the error rate. The fall-back speeds are 16.8 kbits/s, 14.4 kbits/s and 9.6 kbits/s. The effective data transmission rate can be still further increased if data compression is also employed. Some database files can be compressed by as much as 10:1, which gives an effective bit rate as high as 192 000 bits/s.

The V32 terbo *de facto* standard has not yet been ratified as V32 ter by the ITU-T.

V33 Modem

A V33 modem operates at 14.4 kbit/s to provide full-duplex operation over a four-wire presented leased circuit. A technique known as data compression may permit operation at bit rates of up to 24 kbit/s over a conditioned leased circuit. The V33 recommendation includes an optional time-division multiplexer that can divide the 14.4 kbit/s line into subchannels in various multiples of 2.4 kbit/s, e.g. 6×2.4, $2 \times 2.4 + 2 \times 4.8$.

The modulation technique employed is based upon groups of six bits that are trellis encoded to produce a seventh redundant bit which is used for *forward error correction*. The resulting analogue signal has 128 different phase and amplitude positions on the constellation diagram. The fall-back speed is 12 kbit/s with a suitable multiplexing arrangement using five-bit encoding.

V34 Modem

The ITU-T V34 (V FAST) modem standard specifies a maximum bit rate of 28.8 kbits/s. The system uses Trellis-encoding and a variety of signal shaping techniques, such as pre-emphasis and pre-coding, to reduce the error rate to an acceptably low figure. The fall-back rates are 24.4 kbits/s and 19.2 kbits/s.

Short-haul Data Communication

When a short-distance link, within a building, on a multi-building site, or perhaps within a local telephone exchange area, is to be set up between two terminals, some kind of short-haul device can often be employed. The short-haul devices, which are considerably cheaper than conventional modems, are known variously as short-haul, local, or limited-distance modems, as modem eliminators, as baseband modems, and as line drivers, line receivers and line transceivers. The use of line drivers, etc., allows the distance between two terminals to be extended above the 15 m limit imposed by the ITU-T V24 recommendation. A line transceiver IC contains both a line driver and a line receiver, as shown by Fig. 3.5(a). Figure 3.5(b) shows how a EIA 232E data link can be obtained over a distance in excess of 15 m using a line driver and a line receiver.

A modem eliminator is used to connect two synchronous terminals together that are sited within the same office or within the same building at a distance of less than 100 m. A modem eliminator will be able to provide all the necessary synchronization and interface signals and it will replace two synchronous modems.

Fig. 3.5 (a) Line transceiver. (b) EIA 232E using a line driver and receiver.

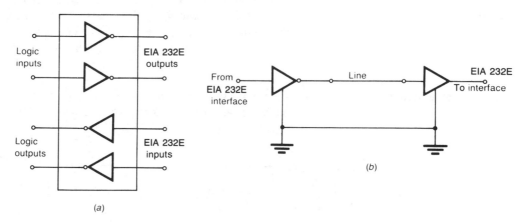

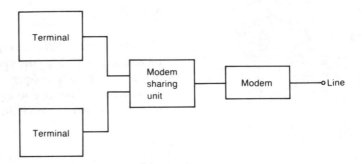

Fig. 3.6 Modem sharing unit.

Modem sharing Units

A modem-sharing unit allows two terminals to share a single modem; the concept is shown in Fig. 3.6. The two terminals contend for access to the modem and the first terminal to call is given the exclusive use of the modem and the telephone line. Once that terminal has transmitted its data and the modem has been released the other terminal is able to contend for the use of the modem and the link. A modem-sharing unit will operate at up to 19.2 kbit/s and is able to pass either synchronous or non-synchronous signals.

Multiplexed Modems

Some higher-speed modems have more than one digital input port and their internal circuitry incorporates a time-division multiplexer. A typical example is shown in Fig. 3.7. The multiplexer allows the four 2.4 kbit/s digital input signals to be combined together to form a composite 9.6 kbit/s data stream. This composite signal modulates the carrier to produce a voice-frequency signal at a baud speed of 2.4 kbauds and this is transmitted over the telephone line. At the distant end of the circuit the v.f. signal is first demodulated and then it is demultiplexed to obtain the four original, distinct 2.4 kbit/s signals. Each of the 2.4 kbit/s signals is passed out of the correct port and thence on to the correct terminal. The use of a multistream modem is cheaper than using four separate sets of 2.4 kbit/s modems and lines.

Operation of a Modem

Before two terminals can send data to one another they must first set up the connections to their respective modems and between each other.

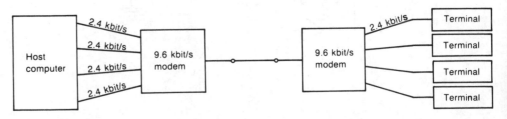

Fig. 3.7 Use of modems with internal multiplexers.

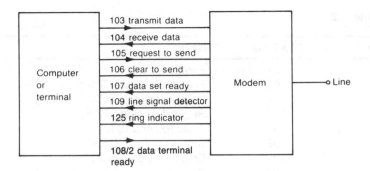

Fig. 3.8 Main parts of the ITU-T V24 and EIA 232 interfaces.

The sequence of events necessary to achieve this is known as a *handshake*.

An interface is the means by which a terminal is connected to a modem that ensures that both electrical and mechanical compatibility are obtained. Mechanical compatibility means that the plugs, sockets, etc., that are used fit one another, and conductors with the same function are connected to the same pins. Electrical compatibility means that the terminal and the modem both use the same voltages to indicate the binary states 1 and 0. The interface consists of a number of circuits that carry the various data and control signals. Specifications for interfaces have been introduced by both the ITU-T and the EIA and these are considered in some detail in Chapter 4. Here only the main components of the ITU-T V24 and EIA 232E interfaces will be considered, and these are shown by Fig. 3.8. The two recommendations are very similar and cover operation at bit rates up to 20 kbit/s.

The ITU-T V24 recommendation provides a set of interchange signals to allow the transfer of serial binary data between a terminal and a modem. The V28 recommendation defines the electrical characteristics of the V24 interface up to 20 kbit/s. The EIA 232E specification gives a complete description of the V24 interface and it is essentially a combination of both V24 and V28 plus anything else that is needed for its implementation. The V24/EIA 232E voltage levels are +3 V to +15 V for the logic 0 level and ON for a control circuit, and −3 V to −15 V for the logic 1 level and OFF for a control circuit; for many modems this voltage range is restricted to ±6 V. Some interface devices use the TTL voltage levels of 0 V and +5 V, and for these a level-shifting device will be necessary. EIA 232E defines 24 circuits as a group and assigns pin numbers to them but it does not specify the actual types of plugs, sockets, etc., to be used. Not all of the 25 circuits specified are normally in use at the same time and the more important circuits are listed in Table 3.2.

Briefly, the functions of these circuits are as follows:

(*a*) 102 provides a reference line from which all the other circuit voltages are measured;

(*b*) data is sent from the terminal to the modem over 103;

Table 3.2

Pin no.	ITU-T no.	Name	Direction
1	101	Earth	
2	103	Transmitted data (TXD)	terminal > modem
3	104	Received data (RXD)	modem > terminal
4	105	Request to send (RTS)	terminal > modem
5	106	Clear to send (CTS)	modem > terminal
6	107	Data set ready (DSR)	modem > terminal
7	102	Common signal return	
8	109	Data channel received line signal detector (CD)	modem > terminal
20	108/1	Connect data set to line	terminal > modem
	or 108/2	Data terminal ready (DTR)	
22	125	Calling indicator (CI)	modem > terminal

(c) data is sent from the modem to the terminal over 104;

(d) a binary 0 or ON signal placed on 105 by the terminal tells the modem that the terminal has data to send;

(e) binary 0 placed on 106 by the modem informs the terminal that it may now transmit its data;

(f) a binary 0 signal placed on 109 by the modem tells the terminal that it is about to receive incoming data;

(g) 107 is used by the modem to inform the terminal when it is operational;

(h) a binary 0 signal from the terminal on 108/1 tells the modem to connect its signal conversion circuitry to the line, this usually happens after a positive voltage has earlier been placed on circuit 125 by the modem to tell the terminal that a calling signal has been received from the line;

(i) the terminal uses 108/2 to inform the modem when it is ready to pass data.

The sequence of events that leads to a successful transfer of data from the transmitting terminal to the receiving terminal is:

(a) the transmitter is activated and it then transmits a bit stream to line;

(b) the distant modem detects the bit stream and uses it to synchronize itself to the transmitter;

(c) the data transfer can then take place; and

(d) the transmitting modem turns itself off after it has made sure that all the transmitted data has been given sufficient time to have arrived at the receiver.

The above sequence is carried out using the V24 circuits listed in Table 3.2. Consider a computer/terminal that is to transfer data to another computer/terminal. The computers/terminals are connected to their respective modems by V24 interfaces and the two modems are connected together via a leased telephone line. When a modem

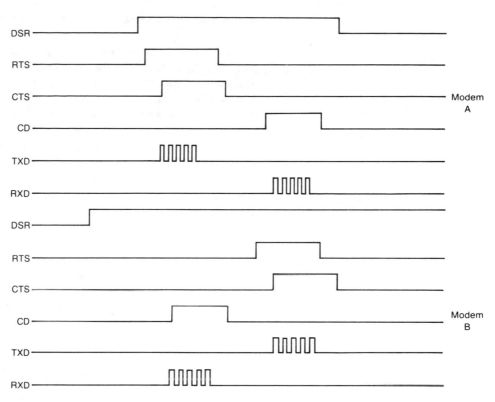

Fig. 3.9 Operation of V24/EIA 232 interface. High voltage = control circuit ON; low voltage = control circuit OFF.

is switched on and is connected to the line the DSR line goes ON and tells the terminal that it may transmit data at any time (see Fig. 3.9). When the terminal has data to transmit it turns the RTS circuit ON and this causes the modem to transmit to line a continuous signal at the carrier frequency. When this signal is received by the distant modem that modem synchronizes itself to the transmitting modem and then signals to its terminal that data is about to be received by turning its CD circuit ON. There is always a certain time delay between the v.f. signal being received from line and the CD circuit turning ON. When the distant terminal is ready to receive the data it will turn its DTR circuit ON. After a short delay, which is longer than the CD delay, the transmitting modem turns its CTS circuit ON and this tells the terminal that it can now transmit its data. The transmitted data TXD passes over circuit 103 to the modem and modulates the carrier to produce a v.f. signal that is then transmitted over the line. At the distant modem the incoming v.f. signal is demodulated and the received data RXD is passed over circuit 104 to the receiving terminal.

When the sending terminal has sent all of its data it turn its RTS line OFF and this causes the transmitting modem to remove the carrier from the line and then turn its CTS circuit OFF. At the distant modem the loss of the carrier wave causes this modem to turn its CD circuit

OFF and this tells the distant terminal that the incoming data has come to an end.

If the circuit is operated on a full-duplex basis this action can take place in both directions at the same time. If, however, half-duplex operation is employed the two modems must *turnaround* before transmission in the opposite direction can occur. At the end of the received data stream the receiving terminal must turn its RTS circuit ON. This will cause the carrier to be transmitted by the called modem and then received by the calling modem when the above procedure will be repeated in the opposite direction, as shown by the lower half of Fig. 3.9. Clearly, there is always some delay before each of the modems can change from its receiving mode to its transmitting mode, or vice versa, and this delay is known as the *turnaround time* (or delay) of the modem. Typically, this delay is of the order of 120 ms per modem.

There are two other circuits, numbers 108/2 and 125, listed in Table 3.2, that have not been mentioned so far. These are the data terminal ready circuit (DTR) and the calling (or ring) indicator. They are used when a modem is operated over a dial-up PSTN connection. When such a connection is to be set up the telephone number of the wanted terminal is dialled — perhaps automatically by the modem — and when the called number answers the modem is connected to the line by a $+V$ signal placed on to circuit 108/2. This signal may be placed by the modem or by an operator pushing a button, or switch, on the modem, the terminal or elsewhere. The DTR circuit of the receiving terminal is taken high to tell its associated modem that it is ready to receive data. When an incoming call is received by the modem, the modem is connected to the line and circuit DSR is turned ON and then the modem is ready to receive the data.

The many IC modems and modem modules that are available use some of the V24 circuits, and, as an example, consider the Rockwell R1212 V22 modem. This uses the TTL logic levels 0 V and +5 V and the following control signals:- $\overline{RTS}$, $\overline{CTS}$, TXD, TDCLK, XTCLK, RLSD, RDCLK, $\overline{DTR}$, $\overline{DSR}$ and $\overline{RI}$. The bar indicates that the pin is active low. TDCLK and RDCLK are the transmit and receive clocks and they are not necessary for non-synchronous communication. XTCLK allows an external clock to be supplied by the user when synchronous transmissions are used. $\overline{RI}$, the ring indicator (the same as CI), goes low to indicate the presence of a ringing signal on the line.

4 Interfaces and Interface Chips

The designer of a system for linking a computer to its peripherals and to other equipments, such as modems and VDUs, has the choice of a large number of different items offered by many various manufacturers. Whatever the choice that is made the various equipments must be able to work with one another. This means that standard interfaces between computers, terminals, peripherals and modems are necessary to ensure compatibility. In this chapter 'terminal' will be used to refer both to computers and to data terminals. The interface between a terminal and a modem must be able to (*a*) transfer data in both directions, (*b*) control the data flow, (*c*) pass clock signals and, perhaps, (*d*) select the bit rate at which data is transmitted over the telephone line. A standard interface might specify the following parameters: (*a*) the physical dimensions of the connectors and the number and layout of their pins; (*b*) the electrical signals that will be applied to each of the pins and their meanings; and (*c*) a functional description of the interface circuits. The ITU-T and the EIA have both developed specifications to standardize the serial interface between a terminal and a modem and/or a peripheral. Parallel interfaces have not been standardized but the most commonly employed is known as the Centronics interface.

The ITU-T standard serial interface is defined by the V24 recommendations and, for most purposes, this is functionally equivalent to the EIA 232 standard. Both of these standards define a set of interchange circuits over which the terminal and the modem can communicate, either synchronously or non-synchronously. Each circuit has a particular function and it is active when the binary 0 voltage level (a positive voltage) is applied to it by either the terminal or the modem. The V24 standard defines more circuit functions, 55 as opposed to 22, than does EIA 232 but in the main the two standards are compatible. V24 is only used for terminal-modem interfaces but EIA 232 is also used for linking peripherals to a terminal and for connecting data equipments within the same building when a line driver and a line receiver would be used rather than a modem. The electrical characteristics of the interface, such as resistances and signal levels, are defined in another ITU-T recommendation, the V28, but they are an inherent part of EIA 232. EIA 232 allows the use of voltages that are sometimes inconveniently high and its bit rate—distance product is rather limited. Newer standards, the EIA 422/423/449/562 and 485, have therefore been introduced but the EIA 232 is still widely employed.

Both the ITU-T and EIA recommendations assume that a system consists of a link like that shown in Fig. 4.1, where one, or perhaps both, of the terminals is a computer. The computer is known as the *data terminal equipment* (DTE), and the modem is known as the *data communication equipment* (DCE).

Very often, however, a link consists merely of one terminal, perhaps a computer, connected to another terminal, a VDU or a printer perhaps, and it is then not clear which terminal is to be regarded as the terminal and which as the modem. A circuit known as a *null modem* will have to be used.

There are two basic methods of transferring data over a serial interface. The first method uses unbalanced conductors, as shown in Fig. 4.2(*a*), in which only one signal conductor is used to transmit data. A digital signal sent from the terminal to the modem takes the transmit data conductor high or low, depending on the logic level of the signal, and at the modem the signal is measured relative to the common signal earth return conductor. The alternative method, known as balanced or differential operation and shown in Fig. 4.2(*b*), uses two signal conductors for each direction of transmission. A positive voltage applied by the terminal to one conductor with respect to the other conductor indicates logical 0 while a negative voltage applied

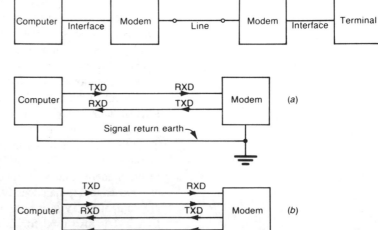

Fig. 4.1 Data circuit for V24/EIA 232 recommendations.

Fig. 4.2 (*a*) Use of unbalanced conductors and (*b*) use of balanced conductors for serial transmission.

to the same conductor transmits logical 1. The modem measures the difference in voltage between its two receive conductors to determine the received signal. If there has been any interference or noise it will have affected both conductors similarly and so the voltage difference is affected to a much lesser extent than if a single conductor is employed. Because differential operation largely overcomes the adverse effects of noise and earth potentials it can be used for higher bit rates and/or longer distances than can unbalanced operation.

Centronics Parallel Interface

A parallel interface is employed for the connection of a peripheral, such as a printer or an X−Y plotter, to a computer. Some form of handshaking is always necessary because the maximum rate at which a printer can receive data and print it out is always less than the rate at which a computer can transmit it. The most commonly employed parallel interface is the *Centronics interface* and this provides an eight-bit data bus plus a number of handshaking and printer control lines. The Centronics interface employs a 36-pin plug; pins 2 through to 9 are the eight data pins D_0 through to D_7, pins 1, 10 and 11 are the three handshaking lines, and pins 13 to 17 may be used for various printer control purposes. Pins 19 to 30 and pin 32 are earth return lines and the remaining pins are either not connected or carry power supply voltages.

When the computer has data to transmit to the printer it first checks the state of the busy line (pin 10); if this is low the printer is able to receive data. The computer will place the first word on to the eight data pins and take the strobe line (pin 1) low. The data will then be fed into the printer's buffer store where it is temporarily held before it is printed. The printer takes the acknowledge line (pin 11) low to inform the computer that the data word has been received and then the computer places the next word on to the data pins. The data is fed word by word into the buffer store at a rate faster than the printer can print it and so the buffer is soon full and unable to take more data. When the buffer is full the busy line is taken high and this tells the computer to stop sending more data. When the printer is again able to receive data, usually when the buffer store is empty, the acknowledge and the busy line are both taken low and the computer resumes sending its data to the printer. The busy line is also taken high if the printer is inoperative for some reason. Not all Centronics interfaces make use of the busy line, some just rely upon the acknowledge line to tell the computer when it can send data.

The other signals which may be used in a Centronics interface are paper exhausted PE (pin 12), autofeed (pin 14), int (pin 31), which is used to reset the printer and clear its buffer, and error (pin 15).

V24 Interface Circuits

The ITU-T V24 recommendations specify 55 interchange circuits in all, for serial binary data transfer but no interface would ever use all of them. Any practical interface only uses a sub-set of these 55 circuits with just fifteen of them being commonly employed. The V24 recommendations do not attempt to define the electrical characteristics of the circuits but leaves this task to the V28 recommendations. Mechanical details are also left unspecified. Two sets of circuits are listed: one set, numbered in the range 100 to 199, is concerned with control, data and timing circuits; the other set, numbered in the range 200 to 299, deals with automatic dialling circuits.

Most of the the 55 interchange circuits are listed, in numerical order, in Table 4.1. Those omitted are three variations of 102 and 12 auto-dialling circuits. The most important, and hence most commonly employed, interchange circuits are numbers 101, 102, 103, 104, 105, 106, 107, 108/1 or 108/2, 109 and 125 and their action has already been outlined in Chapter 3. The calling indicator is used in some automatic answering modems to indicate that an incoming call is being received. The circuit turns ON while the ringing current is being received; the terminal can then respond by taking circuit 108/1 high to connect the modem to the line. The circuit 106 is often called clear-to-send (CTS). This is actually the name given to the corresponding EIA 232 circuit but it is often thought to be clearer in its meaning.

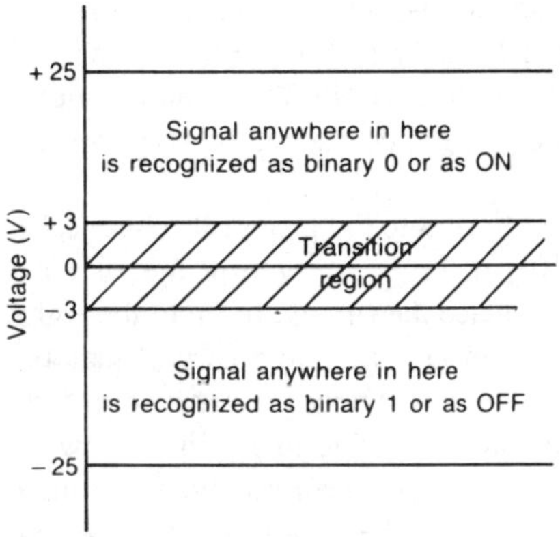

The electrical characteristics of the V24 interface are defined by the V28 recommendations. These specify that unbalanced circuits are used with a positive voltage of between 3 and 25 V representing ON for a control circuit and binary 0 for a data circuit. Similarly, -3 to -25 V represent either OFF or binary 1; this is shown by Fig. 4.3. All the voltages are with respect to the signal earth conductor. The maximum bit rate permitted is 20 kbit/s for a distance of up to 15 m. The V28 recommendation was developed some time ago and its electrical characteristics are not compatible with IC technology, also it has a limited speed−distance product. Two newer standards, originally introduced as EIA 423 and EIA 422 have now been adopted by the ITU-T as the V10 and V11 recommendations.

Fig. 4.3 V24/EIA 232 voltage ranges.

EIA 232 Interface

Probably the most commonly employed serial data interface is the EIA 232 which is really a combination of the most popular circuits of V24, plus the V28 electrical specifications plus whatever else is needed for its practical implementation at bit rates of up to 20 kbit/s. For example, it includes a definition of the pin connections to be used with a standard 25-pin connector although usually fewer than 25 circuits are used. The EIA 232 interface was first introduced in 1962 and has since progressed through various versions to EIA 322 D in 1987 and EIA 232 E in 1994. Table 4.2 gives the main parameters of the EIA 232 E serial interface standard.

Table 4.3 gives details of the EIA 232 interface and their ITU-T V24 pin equivalents.

Briefly, the operation of the EIA 232 interface is as follows.

Table 4.1 ITU-T V24

Circuit number	Circuit name	Direction of signal	Type of circuit
101	Earth	— —	—
102	Signal return earth	terminal > modem	—
103	Transmitted data (TXD)	terminal > modem	Data
104	Received data (RXD)	modem > terminal	Data
105	Request to send (RTS)	terminal > modem	Control
106	Ready for sending (RFS)	modem > terminal	Control
107	Data set ready (DSR)	modem > terminal	Control
108/1	Connect data set to line	terminal > modem	Control
108/2	Data terminal ready (DTR)	terminal > modem	Control
109	Data channel received line signal detector (DCD)	modem > terminal	Control
110	Data signal quality detector	modem > terminal	Control
111	Data signal rate selector	terminal > modem	Control
112	Data signal rate selector	modem > terminal	Control
113	Transmitter signal element timing	terminal > modem	Timing
114	Transmitter signal element timing	modem > terminal	Timing
115	Receiver signal element timing	modem > terminal	Timing
116	Select standby	terminal > modem	Control
117	Standby indicator	modem > terminal	Control
118	Transmitted secondary channel data	terminal > modem	Data
119	Received secondary channel data	modem > terminal	Data
120	Transmit secondary channel line signal	terminal > modem	Control
121	Secondary channel ready	modem > terminal	Control
122	Secondary channel received line signal detector	modem > terminal	Control
123	Secondary channel signal quality detector	modem > terminal	Control
124	Select frequency groups	terminal > modem	Control
125	Calling indicator (CI)	modem > terminal	Control
126	Select transmit frequency	terminal > modem	Control
127	Select receive frequency	terminal > modem	Control
128	Receiver signal element timing	terminal > modem	Timing
129	Request to receive	terminal > modem	Control
130	Transmit backward tone	terminal > modem	Control
131	Receive character timing	modem > terminal	Timing
132	Return to non-data mode	terminal > modem	Control
133	Ready for receiving	terminal > modem	Control
134	Received data present	modem > terminal	Control
136	New signal	terminal > modem	Control
140	Logbook test	terminal > modem	Control
141	Local logbook	terminal > modem	Control
142	Test indicator	modem > terminal	Control
191	Transmitted voice answer	terminal > modem	Control
192	Received voice answer	modem > terminal	Control

Dial-up Connection via the PSTN

Refer to Fig 4.4; the connection process starts with the data terminal ready circuit taken ON to tell the modem that the terminal wants to set up a dialled connection via the PSTN. The telephone number of

Table 4.2 EIA 232E

Mode of operation	Single ended
Number of drivers and receivers per line	1 of each
Maximum bit rate	20 kbits/s
Maximum line capacitance	2500 pF
Driver output voltage	± 5 V to ± 15 V
Driver load resistance	3 kΩ to 7 kΩ
Receiver sensitivity	± 3 V
Receiver input resistance	3 kΩ to 7 kΩ

Table 4.3

Pin no.	EIA 232 label	ITU-T V24 equivalent	Name	Direction	Type
1	AA	101	Shield	—	—
2	BA	103	Transmitted data (TXD)	terminal > modem	Data
3	BB	104	Received data (RXD)	modem > terminal	Data
4	CA	105	Request to send (RTS)	terminal > modem	Control
5	CB	106	Clear to send (CTS)	modem > terminal	Control
6	CC	107	Data terminal equipment (DCE)	modem > terminal	Control
7	AB	102	Signal earth return	—	
8	CF	109	Received line signal detector (DCD)	modem > terminal	Control
12	SCF	122	Secondary received line signal detector	modem > terminal	Control
13	SCB	121	Secondary clear to send	modem > terminal	Control
14	SBA	118	Secondary transmitted data	terminal > modem	Data
15	DB	114	Transmitter signal timing	modem > terminal	Timing
16	SSB	119	Secondary received data	modem > terminal	Data
17	DD	115	Received signal timing	modem > terminal	Timing
18	LL		Local loopback	—	Control
19	SCA	120	Secondary request to send	terminal > modem	Control
20	CD	108/2	Data terminal ready (DTR)	terminal > modem	Control
21	RL	110	Remote loopback	modem > terminal	Control
22	CE	125	Ring indicator (RI)	modem > terminal	Control
23	CH	111	Data signal rate detector	terminal > modem	Control
24	DA	113	Transmitted signal timing	terminal > modem	Timing
25	TM	112	Test mode	modem > terminal	Control

the distant terminal is then dialled by the modem; the number may already be stored in the modem or it may be supplied to it via the transmitted data line. When the call gets through to the called terminal the ring indicator (RI) of the distant modem is turned ON to inform the called terminal of the incoming call. If the called terminal is ready to receive data its data terminal ready (DTR) circuit will be turned ON; this will cause the called modem to connect to the PSTN call. The called modem also turns ON its data terminal equipment (DCE) circuit to inform the called terminal that it is ready to receive the

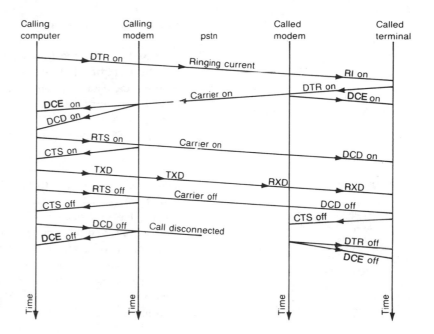

Fig. 4.4 Operation of the EIA 232 interface for a dial-up connection.

incoming data and transmits a carrier back to the calling terminal to indicate that the call has been answered. This carrier turns ON the received line signal detector (DCD) in the calling modem and then the calling modem will turn ON its data terminal equipment (DCE) circuit to indicate to the calling terminal that the line connection has been made. When the calling terminal is ready to start sending its data it turns its request to send (RTS) circuit ON and the calling modem responds by turning its clear to send (CTS) circuit ON and transmits a carrier over the circuit to the called modem. This turns ON the received line signal detector (DCD) in the called modem. The calling terminal is now able to transmit its data and the called modem is able to receive the incoming data.

If the called terminal has data to send to the calling terminal it will turn its request to send (RTS) circuit ON and the same procedure will be repeated in the reverse direction. If the connection is full-duplex the request to send (RTS), and received line signal detector (DCD) circuits will be held ON at both ends of the circuit. Both modems will keep their clear to send (CTS) circuits ON and allow both modems to transmit data at the same time. When all the data has been transmitted the calling terminal turns its request to send (RTS) circuit OFF and the calling modem responds by turning its clear to send circuit OFF. This, in turn, causes the received line signal detector (DCD) to be turned OFF which disconnects the modem from the telephone line and also turns OFF the data terminal equipment (DCE) circuit. At the other end of the line the loss of the incoming carrier causes the called modem to clear down also.

Leased Line Connection

A leased line connection does not have any protocol for the establishment of a link before data transmission commences, but at all times the data terminal equipment (DCE) circuit must be turned ON. When a terminal is ready to transmit data it turns ON its request to send (RTS) circuit and the modem responds by turning ON its clear to send (CTS) circuit. The calling terminal can then transmit its data.

Null Modem

Both ITU-T V24 and EIA 232 always refer to a data terminal equipment (DTE) connected to a data communications equipment (DCE); this is satisfactory when dealing with a computer or a terminal that is connected to a modem. Sometimes, however, a data circuit may consist of a terminal directly connected to a computer; the terminal could then be considered to be either a DTE or a DCE. Similarly, if a terminal is connected to a printer either could be regarded to be the DTE or the DCE. A computer or terminal configured as a DTE transmits data on pin 2 and receives data on pin 3; but if it is configured as a DCE it transmits data on pin 3 and receives data on pin 2. The problem is overcome by the use of a *null modem*. A null modem consists of a length of cable with a connector at each end which interchanges the transmit data and receive lines as well as all the necessary control circuits. The connections of a null modem are shown by Fig. 4.5.

An alternative null modem, which is often used in conjunction with PCs, is shown by Fig. 4.6.

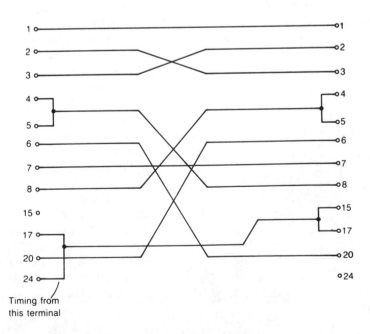

Timing from
this terminal

Fig. 4.5 Null modem.

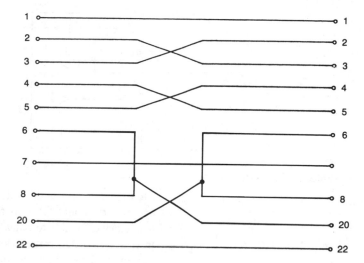

Fig. 4.6 Alternative null modem.

EIA 422

The EIA 422 specification is equivalent to the ITU-T V11 recommendation and it defines the electrical characteristics of a balanced digital interface circuit. It does not define the interface signals, the mechanical characteristics or the method of data transfer. It specifies that line drivers and line receivers are used for bit rates of up to 10 Mbit/s over a maximum distance of 12 m, and of 100 kbit/s up to a maximum distance of 1200 m. The line driver ought to be able to transmit a minimum 2 V differential signal to a two-wire line that is terminated in a resistance of 1000 Ω. The line receiver is required to be able to detect a ±200 mV differential signal in the presence of a noise or interference signal of up to ±7 V. Figure 4.7(a) shows how an EIA 422 link is set up; the line driver could be, for example, a Texas SN 75158/9 and the line receiver a Texas μA9637A.

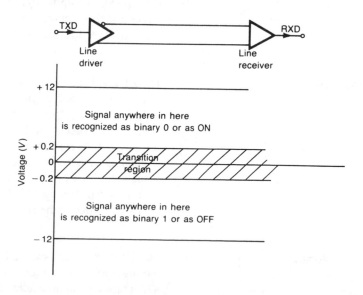

Fig. 4.7 (a) EIA 422 link. (b) EIA 422 voltage ranges.

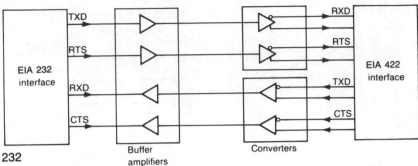

Fig. 4.8 Conversion of an EIA 232 interface into an EIA 422 interface.

Figure 4.7(*b*) shows the voltages which represent binary 0 and ON, and binary 1 and OFF in an EIA 422 interface. It is sometimes necessary to be able to convert an EIA 232 interface into an EIA 422 interface in order to use a longer length of cable and Fig. 4.8 gives one method by which this may be done.

EIA 423

The EIA 423 specification is equivalent to the ITU-T recommendation V10 and it defines the electrical characteristics of an unbalanced digital interface circuit. Like EIA 422 it does not specify the signals, the connector pins or the method of data transfer. It can provide a longer distance circuit than can EIA 232 because it employs driver voltages of only 4 V to 6 V instead of 3 V to 25 V together with much greater receiver sensitivity of ± 200 mV as opposed to ± 3 V. Further features of the EIA 423 specification are that the driver outputs go into a high-impedance state whenever the power is off, and the transmitted data signals are waveshaped to reduce crosstalk and radiation from the conductor. The basic arrangement of an EIA 423 interface is the same as that of EIA 232 and it can provide a bit rate of up to 100 kbit/s for distances of up to 10 m and of 1 kbit/s up to 1200 m. The line driver and line receiver could, once again, be Texas devices such as the μA9636A and the μA9637A, respectively.

EIA 449 Interface

The EIA 449 specification defines the signals, the connectors, etc., that should be used in conjunction with both EIA 422 and EIA 423. It defines 30 interface circuits and their operation at bit rates up to 2 Mbit/s and details a 37-pin connector with an additional 9-pin connector if a secondary channel is used. The majority of the defined circuits correspond with EIA 232/V24 circuits as can be seen from Table 4.4. Usually only a few of these circuits are used. For operation over a short distance EIA 422 transmission will be adequate, but for higher performance the circuit will need to be operated using EIA 423.

EIA 485 Interface

The EIA 485 specification is based upon EIA 422 but it has been modified to allow a multi-point interface to be employed. The EIA 422 interface cannot handle other than point-to-point interfaces since

Table 4.4

	EIA 449	EIA 232 pin		EIA 449	EIA 232 pin
SG	Signal ground	7	SQ	Signal quality	21
SC	Send common	—	NS	New signal	—
RC	Receive common	—	SF	Select frequency	—
IS	Terminal in service	—	SR	Signalling rate detector	23
IC	Incoming call	22	SI	Signalling rate indicator	—
TR	Terminal ready	20	SSD	Secondary send data	14
DM	Data mode	6	SRD	Secondary receive data	16
SD	Send data	2	SRS	Secondary request to send	19
RD	Receive data	3	SCS	Secondary clear to send	13
TT	Terminal timing	24	SRR	Secondary receiver ready	12
ST	Send timing	15	LL	Local loopback	—
RT	Receive timing	17	RL	Remote loopback	—
RS	Request to send	4	TM	Test mode	—
CS	Clear to send	5	SB	Standby indicator	—
RR	Receiver ready	8			

if more than one line driver connected to a bus is simultaneously enabled an excessively large current may flow from the drivers which could damage the ICs. EIA 485 has been designed to allow up to 32 line drivers and line receivers to be connected to a common bus and at the same time satisfy all the requirements of EIA 422. A typical arrangement for an EIA 485 multi-point interface is shown by Fig. 4.9.

EIA 562 Interface

Increasingly, portable equipments such as notebook computers are using 3 V power supplies because of the consequent power savings.

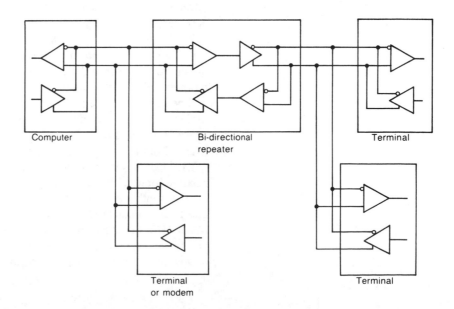

Fig. 4.9 EIA 485 multi-point interface.

Table 4.5 EIA 562

Mode of operation	Single ended
Number of drivers and receivers per line	1 of each
Maximum bit rate	64 kbits/s
Maximum line capacitance	2500 pF for up to 20 kbits/s
	1000 pF for 20 to 64 kbits/s
Driver output voltage	± 3.7 V to ± 13.2 V
Driver load resistance	3 kΩ to 7 kΩ
Receiver sensitivity	± 3 V
Receiver input resistance	3 kΩ to 7 kΩ

The EIA 562 serial interface standard has been introduced for use with portable and low-voltage equipments and it is electrically compatible with EIA 232 E. The parameters of the EIA 562 standard are listed in Table 4.5.

The driver output voltage swings differ from EIA 232 E but the receiver sensitivities are the same. The ± 3.7 V minimum output voltage swing allows a 562 device to communicate with a 232 device which has a sensitivity of ± 3 V, although the noise margin is then a mere 0.7 V. The power dissipation of a 562 driver is just 0.55 times that of a 232 E driver because of the reduced output voltage.

Interface Chips

Within a microprocessor, or a computer, data is moved around in parallel form using either a 16-bit or a 32-bit bus. It is not normally possible to conncct a pcripheral equipment directly to the microprocessor data bus because it will almost certainly have one, or more, of the following operating differences. (*a*) They might have different voltages and/or currents; electrical buffering will be necessary if the peripheral operates at a different voltage (e.g. EIA 232's ± 3 to ± 12 V as opposed to TTL's $+5$ V and 0 V) or requires a greater current than the microprocessor is able to supply. (*b*) Usually, the microprocessor and the peripheral will operate at very different bit rates and handshaking lines to control the timing of data transfers will be necessary. Frequently, a buffer store is employed to provide timing control. The microprocessor transmits data into the buffer where it is stored until the peripheral is ready to process it and then it is fed out at the bit rate of the peripheral.

The input, or output, transfer of data may be carried out using either a parallel-interface or a serial-interface IC. Parallel-interface ICs are known as either a peripheral-interface adaptor (PIA), or as a versatile interface adaptor (VIA). Examples of the former are the Motorola MC6821/2, the Motorola MC68488 and the Rockwell 6522. A PIA has 16 input/output port lines and each of them can be configured to operate as either an input port line or an output port line. These

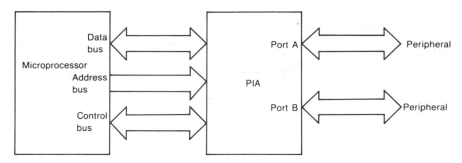

Fig. 4.10 Basic operation of a PIA.

port lines are labelled as PA_0 to PA_7 and as PB_0 to PB_7. The basic operation of a PIA is illustrated by Fig. 4.10 in which it has been assumed that only eight lines are used to connect the PIA to the peripheral. Parallel interfaces are not standardized but very often the *Centronics interface* is employed.

The parallel transmission of data signals cannot be employed over distances greater than about 2 m, and so for greater distances an IC that can also provide both parallel-to-serial conversion and serial-to-parallel conversion is required. Such ICs are known by various names and provide a variety of circuit functions. Examples include (*a*) the universal asynchronous receiver/transmitter (UART), e.g. the Motorola MC68681, (*b*) the universal synchronous/asynchronous receiver/transmitter (USART), e.g. the Intel 8251A, (*c*) asynchronous communication interface adaptor (ACIA), e.g. the Motorola 6850 or the Rockwell 6551, and (*d*) serial communication controller (SCC), e.g. the AMD Z85C30. The microprocessor is connected to the ACIA, or other IC, by eight data lines and a number of control lines, and the ACIA is connected to a terminal by a standard serial interface, such as EIA 232 or V24. The principle of operation of a PIA is illustrated by Figs 4.11(*a*) and (*b*).

With most microprocessors the interface IC is assigned an address in the memory map so that the chip can be programmed to act in the required manner by using the appropriate instructions. If, as for example with the Zilog Z80 microprocessor, the interface IC is not memory mapped but is assigned a separate input/output address, access to it is obtained by the use of IN and OUT instructions.

The description of the operation of a PIA or an ACIA, or any other interface chip, requires a reference to a particular device and in this chapter the Motorola MC6821 PIA and the Motorola 6850 ACIA have been selected. Although these two ICs were designed for use with Motorola's own range of microprocessors they can also be used with many other makes of microprocessor. The two devices are cheap, readily available, with an adequate performance for many applications, and, most important for the purposes of this chapter, they are amongst the easiest to understand, program and use. Today, the device most often used with PCs is the 16550A UART.

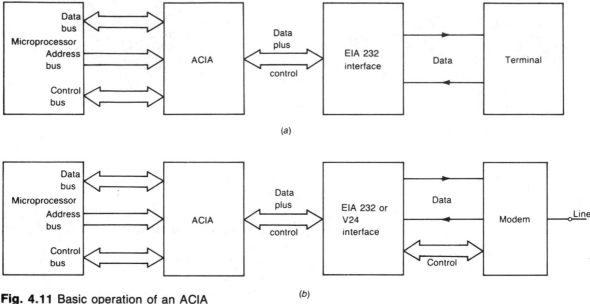

Fig. 4.11 Basic operation of an ACIA connected (*a*) to a local terminal and (*b*) to a modem.

Peripheral Interface Adaptor

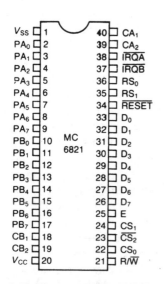

Fig. 4.12 Pin connections of the MC6821 PIA (*Courtesy Motorola Ltd*).

A peripheral interface adaptor (PIA) is an IC whose function is to control the transfer of parallel data between a microprocessor and a peripheral device. The pin connections of the MC6821 PIA are given by Fig. 4.12 and the basic internal block diagram of the IC is shown by Fig. 4.13. Three of the PIA registers can be programmed by the microprocessor. (*a*) The control register is used to control the operation of the PIA. (*b*) The data direction register is used to determine the direction, input or output, in which data flows through each of the input/output ports. If a bit in the data direction register is loaded with binary 0 the corresponding bit in the peripheral interface becomes an input, while if the direction register is loaded with binary 1 the corresponding bit in the peripheral interface becomes an output. This is shown by Fig. 4.14. The output register is used to hold the data that is to be transferred into, or out of, each port. This register links the data bus lines D_0 to D_7 to the input/output peripheral interface port lines PA_0 to PA_7, or PB_0 to PB_7. Each of these registers is given a set of memory locations in the memory map of the microprocessor and this enables the microprocessor to write to, or read from, any of the registers.

The function of each pin of the 6821 is as follows.

(*a*) Pins 2 to 9, and pins 10 to 16, are two groups of bi-directional data lines that can be programmed to act as either inputs or outputs.

(*b*) Pins 18, 19, 39, and 40 are the control lines for ports PA and PB, respectively. Their operation is described later.

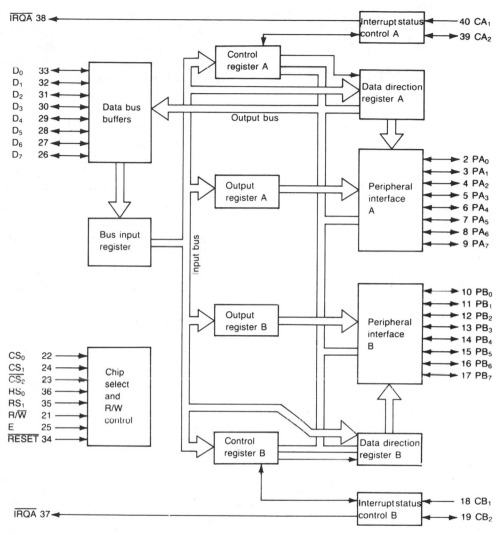

Fig. 4.13 Block diagram of the MC6821 PIA (*Courtesy Motorola Ltd*).

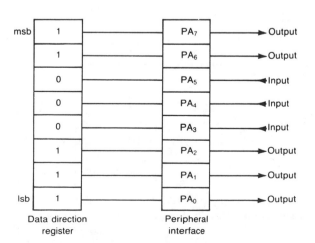

Fig. 4.14 Programming the peripheral interface.

(c) Pin 21 is the read/$\overline{\text{write}}$ line that is used by the micro-processor to control the direction of data transfers over the data bus. If the R/$\overline{\text{W}}$ pin is taken low data will be written by the microprocessor into the PIA (assuming that the PIA has already been selected and enabled). When the R/$\overline{\text{W}}$ line is high the PIA is set up for a transfer of data to, or from, the data bus.

(d) Pins 22, 23 and 24 are the chip-select lines that are used by the microprocessor to select the PIA. Pins CS_0 and CS_1 must be high and pin $\overline{CS_2}$ must be low for the PIA to be selected. Once the PIA has been selected the transfer of data is controlled by the R/$\overline{\text{W}}$ and enable pins.

(e) Pins 35 and 36 are the register select lines RS_0 and RS_1 and signals applied to these pins determine which of the six internal registers is to be selected. The data direction register A (or B) and the output register A (or B) have the same address in the memory map of the microprocessor and the two are distinguished between by means of the bit stored in position 2 of the control register, see Table 4.5. The two register select pins are connected to the two lowest-numbered address pins of the microprocessor.

If the bit 2 of the control register is at the logic 1 level the output register is accessible for programming, but if the bit CRA_2 is at the logical 0 level the data direction register is accessed. The pins given in Table 4.6 must be connected to the address bus of the microprocessor and Fig. 4.15 shows

Table 4.6

RS_1	RS_2	CRA_2	CRB_2	Register selected
0	0	1	X	Output register A Note: X = 'don't care'
0	0	0	X	Data direction register A
0	1	X	X	Control register A
1	0	X	1	Output register B
1	0	X	0	Data direction register B
1	1	X	X	Control register B

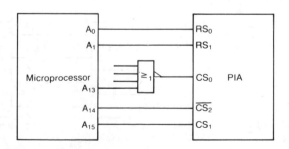

Fig. 4.15 Addressing the MC6821 PIA.

Table 4.7

Address	Register	CS_2	CS_1	CS_0	RS_1	RS_1
$8000	Data direction/output A	0	1	0	0	0
$8001	Control A	0	1	0	0	1
$8010	Data direction B	0	1	0	1	0
$8011	Control B	0	1	0	1	1

a possible arrangement. The addresses of the internal registers will be four consecutive locations in the memory map of the microprocessor. If the base address is $8000 it will be that of both the A data direction register and the A output register. $8000 is the hex equivalent of 1000 0000 0000 0000 and hence A^{15} should be connected to CS_1, A^{14} should be connected to $\overline{CS_2}$, A^{13} should be connected to CS_0 but only after it has been inverted. The logical states of the register selection pins are given by Table 4.7. Although this arrangement addresses the six registers correctly it does overwrite a lot of the memory space, i.e from $8000 upwards. If this is undesirable the amount of overwrite can be reduced by also decoding a number of other memory lines, between A^2 to A^{12}, by connecting the lines to the input of a NOR gate along with line A^{13}. If all the intermediate address lines are decoded there will be no overwrite at all.

(f) Pin 25 is the enable pin; it may either be used to enable the PIA or it may be used as another chip select line. It receives a clock signal from the microprocessor that provides the timing of the data transfers and it is usually connected to the (ϕ_2) clock pin of the microprocessor.

(g) Pins 26 to 33 are the data transfer lines D_0 to D_7 and they are connected to the corresponding pins on the microprocessor.

(h) Pin 34 is used to reset all the internal PIA registers. It is usually connected to the reset pin of the microprocessor.

(i) Pins 37 and 38 are internal interrupt request lines $\overline{IRQA}$ and $\overline{IRQB}$ and they are the PIA outputs which, when taken low, interrupt the operation of the microprocessor. The two pins may be connected directly to the IRQ pin of the microprocessor. Each interrupt has an internal flag in the control register.

Data Direction Register

The data direction registers are used to determine the direction in which each bit of data is transferred through the PIA. Each register has eight bits and logical 1 in any of them causes the corresponding PA, or PB, port line to act as an output; conversely, logical 0 in any bit makes the corresponding port line act as an input. Before the PIA is programmed it should first be initialized by a system reset and this can be achieved by taking its RESET pin low to clear all registers. Usually, the RESET pin is connected to the RESET pin of the microprocessor. Once the PIA has been configured the microprocessor can send data to the peripheral device by sending the data to the address of the data register.

Output Register

The output register contains the data that is to be transferred into or out of the PIA. It has the same address as the data direction register and whether it or the data direction register is accessed depends upon the logical state of bit 2 in the control register. If this bit is 0 the data direction register is selected; if it is 1 the output register is accessed. This was shown by Table 4.6. This means that after the data direction register has been configured the control register should be loaded with #$04. Very often a data direction register is programmed with either 00 or FF so that an entire byte of data will be transferred in or out. Suppose that port A is to be all inputs and port B is to be all outputs. The programming needed to achieve this is as follows, if the address of port A is $8000/1 and the address of port B is $8002/3. $8000 is for the data direction register/output register and $8001 is for the control register.

LDA A #$00		
STA A $8001	clears bit 2 of control register A and selects data direction register A.	
STA A $8003	clears bit 2 of control register B and selects data direction register B.	
STA A $8000	sets port A all inputs.	
LDA A #$FF		
STA A $8002	sets port B all outputs.	
LDA A #$04		
STA A $8001	resets bit 2 of control register A to select output register A.	
STA A $8003	resets bit 2 of control register B to select output register B.	

Control Register

The function of the control registers is to control the action of the microprocessor side of the IC and to define the mode of operation. The control register A has eight bits, labelled CRA_0 to CRA_7 as shown by Fig. 4.16. Bits CRA_0 and CRA_1 control the control line

Table 4.8

CRA_1	CRA_0	Interrupt input CA_1	Interrupt flag CRA_7	MPU interrupt request $\overline{IRQA}$
0	0	H-to-L active	Set high on H-to-L of CA_1	Disabled: $\overline{IRQ}_2$ stays high
0	1	As above	As above	Goes low when CRA_7 goes high
1	0	L-to-H active	Set high on L-to-H of CA_1	Disabled: IRQ stays high
1	1	As above	As above	Goes low when CRA_7 goes high

CA_1. If a handshaking mode is selected the control lines CA_1 and CA_2 at the PIA output carry handshaking signals. The action of the control lines is shown for the A register by Table 4.8.

The table shows that when the CRA_1 bit is 0 bit 7 is set high whenever there is a high-to-low transition on the CA_1 line. The control lines are usually connected to a peripheral that causes a transition whenever it requires attention from the microprocessor. If CRA_1 is at 1 a low-to-high transition on the CA_1 line sets bit CRA_7. If CRA_1 is at 0 the interrupt line $\overline{IRQA}$ is disabled and $\overline{IRQA}$ stays high. If CRA_0 is at 1 $\overline{IRQA}$ goes low every time the status bit CRA_7 is set. When the PIA is used to interface to a data transmitting/receiving device the CA_1 and CA_2 (and CB_1 and CB_2) lines are employed as handshaking lines. Normally both lines arc sct up as inputs but the CA_2 line can be set up as an output strobe line instead. The handshaking action of these lines is controlled by the control register.

Bit CRA_2 in the control register determines whether the microprocessor addresses the data direction register or the control register and hence the peripheral interface. This is necessary because the two registers have the same address in the memory map of the microprocessor. If CRA_2 is at binary 0 the data direction register is addressed; if the bit is at binary 1 the output register is addressed.

The bits CRA_3, CRA_4 and CRA_5 are used to program the peripheral control lines CA_2 and CB_2. When bit 5 is at logical 0 the control line is programmed as an input interrupt line. The control lines CA_2 and CB_2 have different characteristics when programmed as an output line and this is beyond the scope of this chapter.

The bits CRA_6 and CRA_7 are the interrupt flag bits. When lines CA_2 or CB_2 are programmed as an input line bit 6 is its interrupt flag bit. Bit 7 is set to logical 1 when there is an active transition on the CA_1 or the CB_1 line.

Handshaking

When an external device has data for the microprocessor it must send a *data ready* signal to the PIA. When the PIA has read this byte it will acknowledge receipt by returning an ACK signal back to the

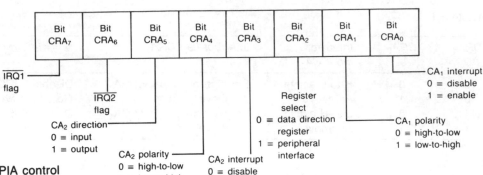

Fig. 4.16 MC6821 PIA control register (*Courtesy Motorola Ltd*).

peripheral. This is known as a handshake; a similar sequence is also used for transmission from the PIA to the peripheral. Consider port A. The handshake line CA_1 has its action governed by bits 0 and 1 in the control register and the action of handshake line CA_2 is determined by bits 4, 5 and 6 of the control register. Refer to Fig. 4.16. When bit CRA_1 is reset to 0 input CA_1 responds to a logical change from 1 to 0. If CRA_1 is at 1 then the CA_1 line responds <u>only</u> to a change from 0 to 1. When the CA_1 line is triggered flag $\overline{IRQ_1}$ is set to 1 to indicate that an input on CA_1 has occurred. This flag is kept up until the microprocessor reads the data register. Once the data register has been read the flag is <u>reset</u> to 0. The action of line CA_2 is very similar except that flag $\overline{IRQ_2}$ is set.

Handshake line CA_2 can be configured as either an input or an output line by the state of bit CRA_5; when it is set as an output line it may be directly set to either the logical 0 or 1 state if bit CRA_4 is set to 1. The state of CA_2 as an output will be the same as the state of CRA_3, so it can always be set by writing into bit 3 of the control register.

When the CB_2 line is used as an output it goes low after the microprocessor has written data into the data register. For this reason the B port lines are usually employed as outputs rather than the A port lines. The handshaking mode, using the B side of the PIA, is illustrated by Fig. 4.17. The control register is first loaded with the word 00100100, i.e. $24. When data is written into the output register the CB_2 line goes low and remains in that state until the peripheral

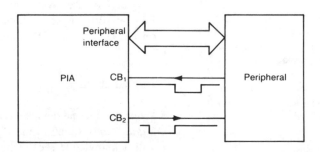

Fig. 4.17 Handshaking.

takes the CB_1 line low to acknowledge receipt of the data. Then CB_2 goes high and this raises the CB_7 flag to indicate to the microprocessor that the data has been accepted and that the next byte of data can now be written into the PIA. Before the microprocessor does this however it must first read the output register in order to clear the CB_7 flag. Suppose the data stored in locations $0800 to $0900 is to be transferred to a peripheral. A possible program, assuming the PIA is initially reset, is then as follows.

```
          LDA A #$FF
          STA A $8002      set B as outputs
          LDA A #$24
          STA A $8003      select output register and handshake
                           mode.
          LDX #$0800
POOL      LDA A 0,X        get data.
          STA A $8003
LOOP      LDA A $8004
          CMP A $A4
          BNE LOOP
          LDA A $8003      reset CRB7
          INX
          CPX #$0900
          BNE POOL
          END
```

Example 4.1

The contents of memory locations $80 and $81 are to be sent to a peripheral. The control register is at address $8001 and the data direction register/output register is at $8000. Write a possible program.

Solution

```
          LDA A #$00       clear accumulator.
          STA A $8001      clear bit 2 of control register A.
          LDA A #$FF
          STA A $8000      port A as outputs.
          LDA A #$04
          STA A $8001      set bit 2 of control register A.
          LDA A $80        load contents of location $80 into the accumulator.
          STA A $8000      output data.
          LDA A $81        load contents of location $81 into the accumulator.
          STA A $8000      output data.
```

Asynchronous Communication Interface Adaptor

An asynchronous communication interface adaptor (ACIA) is an IC that is designed to: (*a*) input data from the microprocessor's parallel data bus, one byte at a time, and serially transmit it bit-by-bit to a non-synchronous terminal or modem, and (*b*) accept serial data

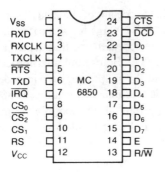

Fig. 4.18 Pin connections of the MC6850 ACIA (*Courtesy Motorola Ltd*).

incoming from a non-synchronous terminal or modem and transmit it, in parallel form, to the data bus of a microprocessor. One of the more commonly employed ACIA ICs is the Motorola 6850 and Fig. 4.18 shows its pinout. The functions of the pins are as follows.

(a) Pins 1 and 12 are for the power supply voltages V_{CC} and V_{SS}.

(b) Pins 2 and 6 are the lines over which the serial data is received and transmitted.

(c) Pins 3 and 4 provide the transmit and receive clocks.

(d) Pins 5, 23 and 24 are the V24/RS 232 signal lines request to send, data carrier detect, and clear to send, respectively, and their uses have been described earlier in this chapter.

(e) Pin 7 is an interrupt request line that is used to interrupt the microprocessor. It therefore signals to the microprocessor when it may receive or send data. The $\overline{IRQ}$ output remains low for as long as the cause of the interrupt is present.

(f) Pins 8, 9, and 10 are three chip select lines. The ACIA is given a specific address in the memory map of the microprocessor and these pins are connected to the address bus of the microprocessor and are used to address the ACIA. The ACIA is selected when the CS_0 and CS_1 pins are high and the $\overline{CS_2}$ pin is low. The pins are connected either directly to the microprocessor address pins as in Fig. 4.19(a), or a decoder may be used as in Fig. 4.19(b). In the latter case it is usual to connect both CS_0 and CS_1 to logical voltage 1 and $\overline{CS_2}$ to the decoder output.

(g) Pins 11 and 13 are the register select and read/write pins respectively. A combination of logic levels applied to these pins, shown by Table 4.9, is used to select one of the four main registers (see Fig. 4.20).

(h) Pin 14 is the enable pin; it must be kept high for the chip select and read/write circuitry to be enabled. The enable signal is usually derived from the ϕ_2 pin of the microprocessor.

(i) Pins 15 to 22 inclusive are eight bi-directional data lines, labelled as D_0 to D_7. These pins are connected to the data pins of the microprocessor.

The simplified block diagram of the 6850 ACIA is shown in Fig. 4.20. Each byte of the data to be transmitted to line is written into the transmit data register by the microprocessor. If the transmit shift register is empty the byte of data is then transferred, in parallel form, from the data register to the shift register. Here start, stop, and perhaps parity bits are added to the data character, and then the data byte is shifted out serially on to the TXD pin. The rate at which data is moved out is determined by an external clock. The microprocessor determines whether the transmit data register is empty either

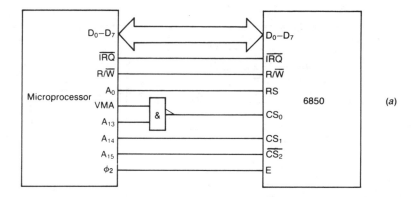

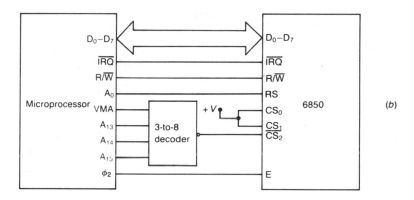

Fig. 4.19 Addressing the MC6850 ACIA.

Table 4.9

$R/\overline{W}$	RS	Action
1	0	Read from status register
1	1	Read from data register
0	0	Write to control register
0	1	Write to transmit data register

by receiving an interrupt request signal, or by first testing the bit in the status register that flags a full register.

Serial data received from the line is passed into the receive shift register. When a complete character has been received its start, stop and any parity bits are removed. The data word is then transferred, in parallel form, into the receive data register from which it can be read out, also in parallel form, by the microprocessor addressing the ACIA and selecting the receive data register. Before each data word is shifted to the receive data register the ACIA must test for three kinds of error. These are: (a) framing error, (b) overrun error, and (c) parity error. This testing is done by checking the appropriate bit

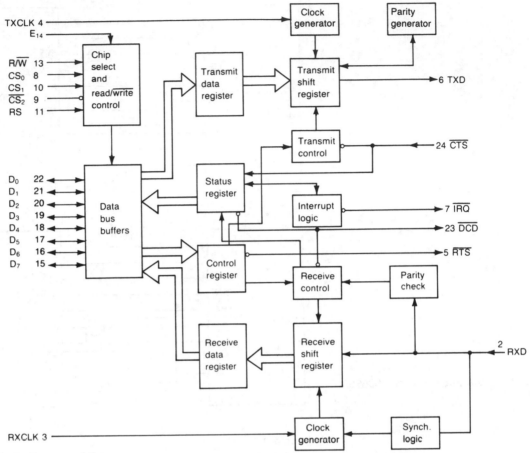

Fig. 4.20 Block diagram of the MC6850 ACIA (*Courtesy Motorola Ltd*).

register. The microprocessor can determine if the receive data register is full by either receiving an interrupt request signal or by checking bit 0 in the status register. If the receive data register is full and there are no errors the microprocessor can then read out the data by storing it in a memory location.

There are *four* basic operations that a microprocessor can perform on the ACIA: these are writing into the control register, reading the status register, writing into the transmit data register, and reading the receive data register. The program instructions for each of these operations assuming that the ACIA has been allocated addresses $B000 and $B001, are:

- (*a*) STA A $B000   write into the control register;
- (*b*) LDA A $B000   read the status register;
- (*c*) STA A $B001   write into the transmit data register; and
- (*d*) LDA A $B001   read the receive data register.

It can be seen that the control and status registers are given the same address as one another and so are the two data registers; this presents no problem, however, since the R/$\overline{\text{W}}$ line determines which of the two registers at each address will be selected.

Control Register

From Table 4.9, the eight-bit control register is selected by RS = 0, R/$\overline{\text{W}}$ = 0 and then signals applied by the microprocessor to the data pins D_0 to D_7 give the conditions shown by Table 4.10.

The counter divide select bits CR_0 and CR_1 allow the frequency of the receiver clock to be divided by either 1, 16, or 64, or to give a master reset of the ACIA, see Table 4.11. The divide-by-1 mode is only used when the receiver clock is synchronized with the received data signals. For non-synchronous operation either $\div 16$ or $\div 64$ must be used.

The bits CR_2, CR_3 and CR_4 are written to when the ACIA is initialized and they determine the number of data bits per word, the number of stop bits inserted at the end of each character, and whether odd or even parity is employed. The details are given by Table 4.12.

Bits CR_5 and CR_6 give *transmit interrupt control* and also control the ready to send line as shown by Table 4.13. When the transmit interrupt is enabled the ACIA will interrupt the microprocessor (if $\overline{\text{IRQ}}$ is connected to the microprocessor) whenever the transmit data

Table 4.10

Data pin	D_0	D_1	D_2	D_3
Control register	CR_0 counter divide select 1	CR_1 counter divide select 2	CR_2 word select 1	CR_3 word select 2
Data pin	D_4	D_5	D_6	D_7
Control register	CR_4 word select 3	CR_5 transmit control 1	CR_6 transmit control 2	CR_7 receive interrupt

Table 4.11

CR_1	CR_0	
0	0	$\div 1$
0	1	$\div 16$
1	0	$\div 64$
1	1	Master reset

Table 4.12

CR_4	CR_3	CR_2	Character	Parity
0	0	0	7 bits: 2 stop bits	Even
0	0	1	7 bits: 2 stop bits	Odd
0	1	0	7 bits: 1 stop bit	Even
0	1	1	7 bits: 1 stop bit	Odd
1	0	0	8 bits: 2 stop bits	
1	0	1	8 bits: 1 stop bit	
1	1	0	8 bits: 1 stop bit	Even
1	1	1	8 bits: 1 stop bit	Odd

Table 4.13

CR_6	CR_5	
0	0	RTS low: transmit interrupt disabled
0	1	RTS low: transmit interrupt enabled
1	0	RTS high: transmit interrupt disabled
1	1	RTS high: transmits constant 0 on TXD; transmit interrupt disabled

buffer is empty. Bit CR_7 is the receive interrupt enable bit. If CR_7 is at the logical level 1 the ACIA interrupts the microprocessor whenever the receive data register is full or if data carrier detect occurs.

Before transmitting data a software master reset is necessary. To achieve this the control register is first programmed with $43 and then it is programmed to give the wanted clock frequency, word length, etc.

Status Register

The different bits of the status register are set, or reset, to indicate the status of the ACIA. The status register is selected by RS = 0 and R/$\overline{W}$ = 1 and the indication of status given by each bit is shown by Table 4.14.

Bit D_0 goes high when a character has been received by the ACIA and the receive data register is full. It is read by the microprocessor to determine when data has been received by the ACIA and is ready to be transferred to the microprocessor.

Bit D_1 goes high to indicate to the microprocessor that the transmit data register is empty and that the microprocessor may transmit the next character to the ACIA.

Bits D_2 and D_3 are V24/RS 232 signals and their use is described on page 66. When $\overline{DCD}$ is high D_2 is set and this takes bit D_1 low. Similarly, if $\overline{CTS}$ is high bit D_3 is set and this also takes bit D_1 low. In both cases bit D_7 is taken low as well.

Table 4.14

Data pin	D_0	D_1	D_2	D_3
Status	Receive data register full	Transmit data register empty	DCD	CTS

Data pin	D_4	D_5	D_6	D_7
Status	Framing error	Receiver over-run	Parity error	Interrupt request

Bit D_4 goes high to indicate that a loss of character synchronization and/or a break in transmission has taken place. It is disabled by bit D_2 going high.

If bit D_5 is high it indicates that a character that has been received from the line has not yet been read from the receive data register and has been lost. This flag is reset by $\overline{DCD}$ going high.

When a parity error has been detected bit D_6 will go high; it can be reset by taking $\overline{DCD}$ high.

Lastly, bit D_7 is the interrupt request that when taken high will interrupt the microprocessor. It is taken high to indicate three different events: (a) the transmit data register is empty, (b) the receive data register is full, and (c) pin $\overline{DCD}$ has gone high to denote that the incoming carrier has been lost.

V24/EIA 232 Interface

When the ACIA is connected to a modem via a V24/EIA 232 interface level translators will be needed to change the TTL voltage levels of the ACIA to the V24/EIA 232 voltages of the modem. This is shown by Fig. 4.21.

Programming the ACIA

When a microprocessor is to transmit data via an ACIA to a distant terminal the ACIA must be programmed to perform in the desired manner. Figure 4.22 gives a flowchart of the steps that must be taken for data to be transferred from the microprocessor to the modem. The ACIA must first be reset and then it must be initialized; this means that the control register must be set up to provide the type of signal that is to be sent to line, i.e. the bit rate, the number of bits per character, and the number of stop bits and the parity employed. When the microprocessor has some data to transmit it first reads the status register and checks whether bit 1 is set or cleared. If it is set the transmit data register is empty and the microprocessor can write the data word into the transmit data register of the ACIA. The ACIA will move the data to the transmit shift register and thence to the TXD pin. If the register is full (bit $1 = 0$) the data cannot yet be transferred to the ACIA and the microprocessor must check bit D_3 to determine whether it must either go back to read the status register again or it must jump to an error subroutine. This procedure continues until all the data has been passed from the microprocessor to the ACIA. A suitable program is as follows. It is assumed that the ACIA addresses are status register $B000, and transmit data register $B001, and the data to be transmitted consists of 7 ASCII characters stored in locations $0051 onwards.

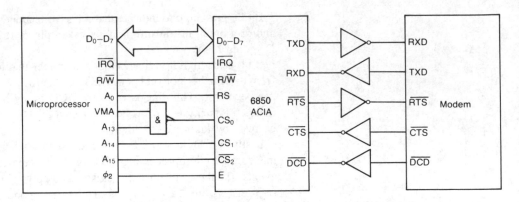

Fig. 4.21 Connection of MC6850 ACIA to a modem.

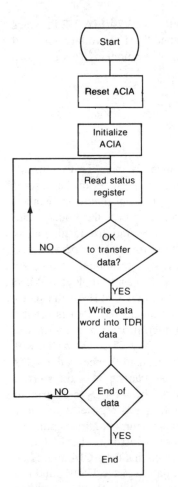

Fig. 4.22 Flowchart for a microprocessor to send data to a ACIA.

	LDA A #$43	
	STA A $B000	reset the ACIA.
	LDA A #$89	
	STA A $B000	put control word into register.
	LDX #$07	
LOOP	LDA $50,X	
	JSR SEND	jump to transmit routine.
	DEX	
	BNE LOOP	
	END	
SEND	LDA B $B000	load accumulator B with contents of status register.
NEXT	ASR B	
	ASR B	move transmit data register empty bit into the carry position.
	BCS TXD	check TDRE bit.
	ASR B	
	ASR B	move CTS bit into the carry position.
	BCC NEXT	
	BSR ERR	go to error routine.
TXD	STA A $B001	data now in the transmit data register.
	RTS	

A routine for the reception of incoming data is now given. The initialization enabled the receiver interrupt, i.e. $CR_7 = 1$, so that the microprocessor can be carrying out other tasks while waiting for data to arrive. The flow diagram for the operation of the ACIA is shown by Fig. 4.23. Whenever the ACIA has a data word in its receive data register it will take its $\overline{\text{IRQ}}$ pin low and this will interrupt the microprocessor. The microprocessor will then execute the receive routine. First the contents of the status register are read and bits 4, 5 and 6 are checked for any error indications, i.e. set at logical 1. If an error is flagged the program will branch into an error subroutine; if no error is flagged the $\overline{\text{DCD}}$ bit must be checked before the data word is read from the receive data register and loaded into memory.

The microprocessor then returns to its main program until the next data word appears in the ACIA. A possible program is as follows.

	PSH A	put contents of accumulator into the stack.
	LDA A $B000	read status register.
	AND #$70	mask three status bits.
	BEQ DATA	if all status bits = 0 get data.
	JSR ERROR	go to the appropriate error subroutine.
	RTI	return to interrupted programme.
DATA	LDA A $B001	
	STA A $0600	store received data in location $0600.
	PUL A	move stack contents back to accumulator.
	RTI	return to interrupted programme.

Communications Software

A computer must be loaded with a communications software package before it is able to communicate with a modem. The serial output port must be configured to match the modem. The following parameters must also be set:

(*a*) The bit rate,
(*b*) The word length, i.e. the number of bits per character,
(*c*) The parity to be employed, i.e. even, odd, or none, and, if non-synchronous transmission is concerned,
(*d*) How many start and stop bits there should be.

Finally, the protocol to be used must be agreed.

Many modems are compatible with the *Hayes command set*. This is a *de facto* standard which has the advantage that it allows a modem to be controlled from the keyboard. For example, to dial a telephone number it is only necessary to type ATD followed by the number. When a Hayes compatible modem is turned on it is automatically started in the local mode. It will then respond to any commands that are entered at the keyboard. When the modem is in the process of setting up a connection via the PSTN, or it is sending or receiving data to/from a distant modem, it is in its on-line mode and then all characters entered at the keyboard are treated as data.

Once the parameters (*a*)−(*d*) have been set up, a communications program must first open the serial port and then go into an endless loop. Each time around the loop the program must (*a*) test to see if data has been entered at the keyboard, and if it has, transmit it, and (*b*) test to see if data has been received at the serial port and if it has, display it on-screen.

A communications software package for a PC should also provide a range of terminal emulations so that it is able to work to many different host computers.

A *Windows* communications package may support *dynamic data exchange* (DDE) which enables the software to pass received data to other Windows programs, and to transmit data generated by the other programs.

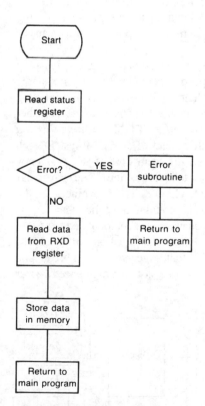

Fig. 4.23 Flowchart for a ACIA to receive data from a modem or terminal.

5 Multiplexing

Long-distance telephone circuits are expensive to use, whether on a leased or a dial-up basis. Often two communicating terminals will not fully utilize the traffic capacity of a link and so some means of increasing the amount of traffic on a line is desirable. There are two main approaches to the problem, known respectively as *multiplexing* and *concentration*. Either technique may incorporate some means of *data compression*.

Multiplexing is the process of combining the data streams originating from a number of separate low-speed data channels to form a single composite high-speed bit stream. The basic concept of multiplexing is illustrated by Fig. 5.1. The *n* data channels can be combined together using either *frequency-division multiplex* (FDM) or *time-division multiplex* (TDM). At the transmitting end of an FDM system each of the data channels is shifted to a different part of the frequency spectrum made available by the telephone line. The particular frequency band allocated to each channel is determined by the frequency of the carrier wave that is modulated by that channel. At the receiving end of the system the channels are demodulated to restore them to their original frequency bands. Applications of the FDM technique are now restricted to V22, 'data-over-voice' and broadband LAN systems. With a TDM system a time slot on a high-speed *bearer*

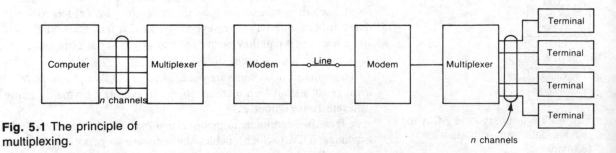

Fig. 5.1 The principle of multiplexing.

is allotted to each of the data channels in time sequence. For the short period of time allotted to it each channel is given the exclusive use of the bearer circuit.

Time-division multiplex uses digital techniques and circuitry which allows it to make full use of the many readily available ICs and it is therefore both cheaper and more reliable than FDM. There is no requirement for any special computer software and a transparent connection is provided between a computer and a terminal. The term 'transparent' here means that the multiplexing process does not affect the flow of data, or the passage of control signals, and so neither the computer nor the terminal is aware that they are not using an exclusive physical link. This means that when a multiplexer is first installed there is no need to alter the software in either the terminal or the computer.

Time-division Multiplex

Time-division multiplex (TDM) is a digital technique that interleaves the data originating from a number of terminals and transmits the aggregate bit stream serially over a higher-speed circuit. At the far end of the link the received data is demultiplexed to separate the channels and then the individual data signals are routed to their correct destinations. The data is transmitted over the bearer circuit in the form of *message frames* that consist of a number of time slots each of which contains data from different channels. The data input by different channels may be combined using either (*a*) bit interleaving or (*b*) character interleaving techniques. Character interleaving is employed for multiplexing non-synchronous channels and bit interleaving is generally used for interleaving synchronous channels.

Character Interleaving

The basic principle of a character-interleaved TDM system is shown by Fig. 5.2. The data generated by each channel is fed serially into the channel store and when the complete character has been received and stored it is shifted in parallel form into a buffer. Usually the start and stop bits are not transfered into the buffer and they are lost. Each character is held in the buffer until the scanning logic connects the store to the high-speed line for one time-slot period. When this happens the character is read out of the buffer at the bit rate of the output line. Once this character has been fed into the line the scanning logic removes the buffer from the line and moves on to the next buffer to enable it to shift out the character that it holds. At the end of this time slot period the scanning logic selects the next channel buffer and connects it to the line and so on for all the channels connected to the multiplexer. When the scanning logic has connected each channel in turn to the bearer circuit it returns to the first channel and this is again enabled to shift out its second character to the line and so on. The

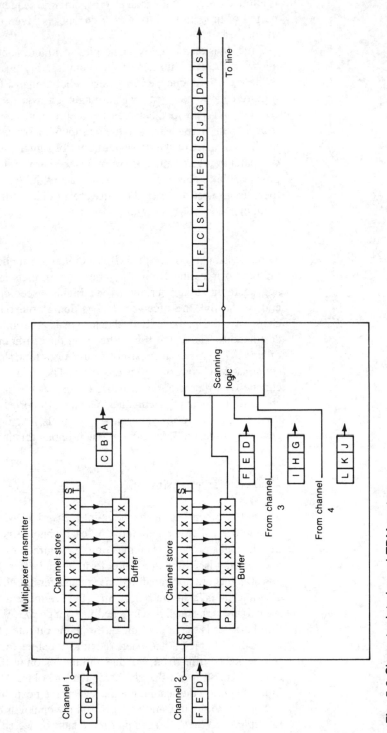

Fig. 5.2 Character-interleaved TDM system.

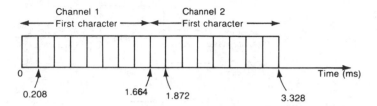

Fig. 5.3 Multiplexer timing diagram.

channel buffer stores introduce a time delay in the transmission of the data equal to between two and three character periods. Suppose, for example, that the four channels shown in Fig. 5.2 each operate at 1200 bit/s using seven-bit ASCII code with one start, one stop, and one parity bit. The time duration of each bit is 1/1200 s or 0.833 ms and each 10-bit character occupies a time of 8.33 ms. Usually, *bit stripping* is employed, which means that the start and stop bits are removed before a character is transmitted (as in Fig. 5.2) over the bearer circuit and are later replaced by the distant multiplexer. Assuming that bit stripping is employed each character will occupy a time slot of 6.67 ms. If the bearer circuit is operated at 4.8 kbit/s each bit transmitted to line will last for 1/4800 s = 0.208 ms. When channel store 1 is connected to the line for 8 × 0.208 = 1.664 ms one stored character is transmitted at the line bit rate of 4.8 kbit/s. At the end of this time channel buffer store 2 is enabled for 1.664 ms and its stored character is sent to line at the 4.8 kbit/s rate, and so on. The timing diagram is shown in Fig. 5.3.

It should be clear that synchronism between the transmitting and receiving clocks is essential so that the received data will be routed to the correct channels. This synchronization is provided by preceding each frame with a synchronizing character; a typical example is 1010101. The receiving terminal looks for the synchronizing character to repeat over several frames and once frame synchronization has been established it continues to monitor the character. The output of a multiplexer is either applied to a modem to convert the composite digital data signal into voice-frequency form suitable for transmission over the telephone network, as in Fig. 5.1, or it is applied to either a Kilostream or a Megastream circuit (see Fig. 5.4).

Figure 5.4(a) shows how the 64 kbit/s speed of the Kilostream circuit can be shared between a number n of lower-speed data channels, while Fig. 5.4(b) shows a Megastream circuit being used to provide 32 × 64 kbit/s channels. Each of the 64 kbit/s time slots can be multiplexed in various ways: the synchronous inputs of between 48 kbit/s to 64 kbit/s are each assigned one time slot; inputs at greater bit rates than 64 kbit/s are allotted 2, 4 or even 8 time slots; and inputs at 19.2 kbit/s and below are serviced two to a time slot. Figure 5.4(c) shows how a Megastream circuit and two multiplexers can be used to provide both data and speech circuits. The multiplexer can multiplex up to 30 speech channels using 64 kbit/s pulse code modulation (PCM) or up to 32 × 64 kbit/s data channels or some combination thereof.

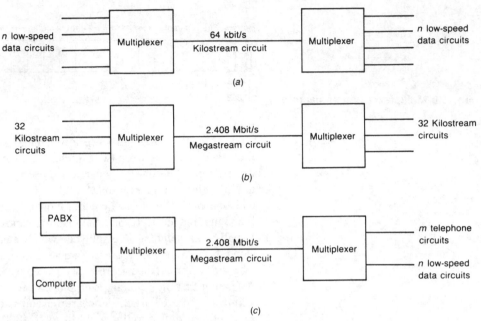

Fig. 5.4 Use of multiplexing on Kilostream and Megastream circuits.

In all three Fig. 5.4 circuits the data channels may be multiplexed further to obtain even more lower-speed channels if required.

Bit Stripping

It has already been mentioned that the start and stop bits of a non-synchronous signal can be removed by the multiplexer before the signal is transmitted to line. *Bit stripping* gives an increase in the effective transmission rate on the bearer circuit. With the ASCII code, for example, the saving in the number of transmitted bits is two in every character or 20%. This saving is, however, reduced by the need to transmit a synchronization character in every frame and the usual saving is somewhere in the region of 10% to 20%.

Example 5.1

Calculate how many 2400 bit/s multiplexed data channels can be transmitted over a 14 400 bit/s bearer circuit if the saving due to bit stripping is 20%. Assume the ASCII code with one start, one stop, and one parity bit.

Solution
Without bit stripping the system can transmit 14 400/2400 = 6 channels.
With bit stripping the system can transmit 6 × 10/8 = 7.5 channels.
Therefore, number of channels = 7.   (*Ans.*)

Different Channel Speeds

Very often the multiplexed channels will not all operate at the same bit rate. When this is the case there are two ways in which the multiplexing can be carried out: (*a*) extra time slots can be allotted to the higher-speed channels — this is known as *variable- channel scanning*; and (*b*) all of the channels are allotted the same time slots and the lower-speed channels leave some of their time slots empty — this is known as *fixed-channel scanning*. The aggregate bit rate on the bearer circuit is equal to the *sum* of the bit rates of the individual channels.

Example 5.2

A 9.6 kbit/s bearer circuit is to carry one 4.8 kbit/s channel plus a number *n* of 1200 bit/s channels using time-division multiplex. Calculate the number of 1200 bit/s channels that can be transmitted.

Solution

$$n = (9600 - 4800)/1200 = 4. \quad (Ans.)$$

The *data transmission efficiency* of a multiplexer is given by

$$\frac{\text{Data transmission}}{\text{efficiency}} = \frac{\text{Effective fast output bit rate}}{\text{Aggregate of slow input bit rates}} \times 100\%$$

$$(5.1)$$

The data transmission efficiency will usually be somewhat greater than 100% for a character-interleaved multiplexer but less than 100% for a bit-interleaved multiplexer.

Example 5.3

A time-division multiplexer has two 4800 bit/s and four 2400 bit/s non-synchronous input channels and an output bit rate of 4.8 kbit/s. If bit stripping is employed calculate the data transmission efficiency.

Solution

Aggregate bit rate = $(2 \times 4800) + (4 \times 2400) = 19\,200$ bits.
The effective output bit rate = $19\,200$ bit/s $\times 10/8 = 24\,000$ bit/s.
Therefore the data transmission efficiency = 125%. (*Ans.*)

Bit Interleaving

A bit-interleaved multiplexer accepts and multiplexes the data from a number of synchronous channels on a bit-by-bit basis. The multiplexed data includes all the synchronizing and control characters generated by the terminals. Each data bit is transmitted by the

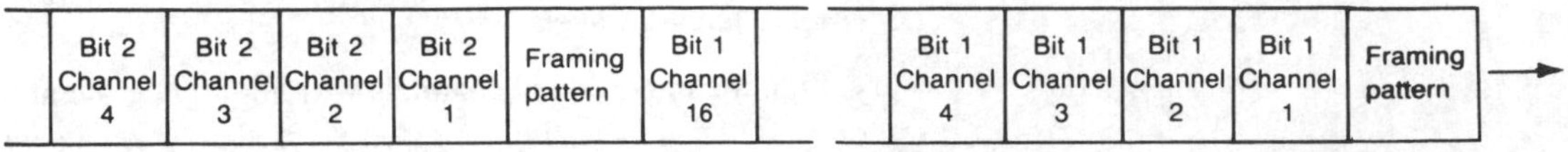

| Bit 2 Channel 4 | Bit 2 Channel 3 | Bit 2 Channel 2 | Bit 2 Channel 1 | Framing pattern | Bit 1 Channel 16 | | Bit 1 Channel 4 | Bit 1 Channel 3 | Bit 1 Channel 2 | Bit 1 Channel 1 | Framing pattern | →|

Fig. 5.5 Bit interleaving.

multiplexer on to the bearer circuit almost as soon as it is received from the terminals, and so large-capacity channel stores are not required and simpler scanning logic can be used. Consequently, the transmission delay is less than for a character-interleaved system, being between two and three bit periods.

Synchronism between the transmitting and the receiving multiplexers is maintained by assembling the individual bits from each terminal into a frame. Each frame will consist of one bit from each of the channels. Since a bit-interleaving multiplexer will transmit all the input bits, including any start and stop bits, it is normally only used for multiplexing synchronous terminals. Each frame is preceded by a bit sequence, known as a *framing pattern*, and this is used to synchronize the distant multiplexer's clock to the transmitter's clock (see Fig. 5.5).

Any errors that occur on the bearer circuit because of line noise and/or interference cannot be corrected by the receiving multiplexer and they will corrupt some of the characters received by some, or all, of the channels. If each of the terminals incorporates its own error-detecting/correcting circuitry this will not matter but many non-synchronous terminals do not possess this capability. This lack of error-correction circuitry tends to limit the possible applications for conventional TDM and it accounts, to some extent, for the increased popularity of *statistical multiplexers*.

Switching Multiplexers

A switching multiplexer allows a single terminal to be able to access more than one computer, and an example of a possible network using these devices is shown in Fig. 5.6. A total of $m+n+p$ terminals are given access to any of three computers. A switching multiplexer transmits its data as separate characters with no addresses appended and so, usually, a separate signalling channel is employed to set up connections.

Statistical Multiplexers

A conventional time-division multiplexer allots time slots on the high-speed bearer circuit to each of its input channels. This practice leads to inefficient utilization of the line if one, or more, of the channels is only intermittently loaded. This problem can be overcome by the use of a *statistical multiplexer* (STDM). The operation of a STDM is based upon the likelihood that at any instant in time some of the terminals will not be transmitting data. Time slots on the high-speed line are dynamically allotted to the active data channels and this

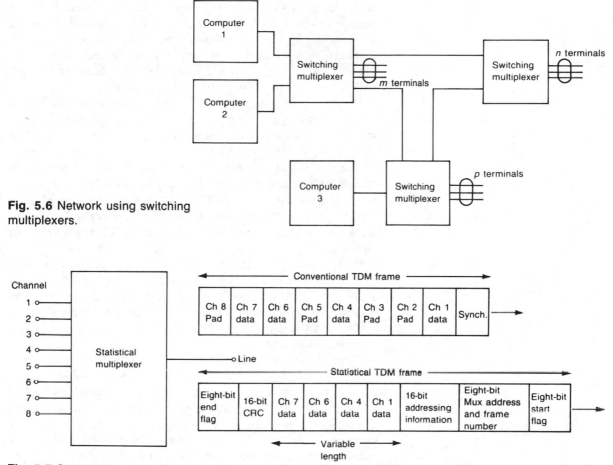

Fig. 5.6 Network using switching multiplexers.

Fig. 5.7 Statistical multiplexing.

increases greatly the efficient usage of the line. An increase in the throughput of data over the bearer circuit, or alternatively, a reduction in the bit rate on the bearer circuit, can be obtained with consequent savings in both line and modem costs. Now the transmission speed on the bearer circuit is less than the sum of the bit rates of the input channels.

The basic concept of statistical time-division multiplexing is shown by Fig. 5.7. Each input data channel is polled in turn by the statistical multiplexer but a time slot is only allotted to those channels that have data ready to transmit. When several inputs are transmitting at the same time it may happen that some of the input data cannot immediately be sent to line since all the output channels are occupied. Data is then queued in buffer stores (flagged to indicate which channel it came from) while the multiplexer is handling data received from other channels earlier. The buffer stores are dynamically allotted to channels as, and when, required. The size of the stores limits the amount of traffic that a multiplexer can handle at one time and if the peak loading is sometimes in excess of this *flow control* will be necessary. The multiplexer has XON and XOFF control characters that are used to defer overflow of the buffer stores. The XOFF character signals to an active terminal that it must suspend its transmission of data for a while and the XON character tells the terminal when it can resume transmitting that data.

The data is transmitted over the bearer circuit in *internodal* frames that are formed by assembling the data from the active channels. Each frame must start with a beginning flag and end with an ending flag to show the start and finish of each frame. Included in each frame are the destination multiplexer address and the frame number, the channel addresses, data, and cyclic redundancy check bits. A typical frame is shown in Fig. 5.7. The CRC bits are a means of error detection and will be discussed in Chapter 9. The channel addresses are necessary to tell the receiving multiplexer which data is for which channel.

The transmission efficiency is the ratio (actual data bits)/(total bits) in a frame. The efficiency can be increased for non-synchronous signals by the use of bit stripping. An increase in efficiency can be obtained with synchronous signals by removing the leading SYN and PAD characters but the SYN character must be replaced by the receiving multiplexer.

There are several different techniques for the formation of the internodal frames: (*a*) Each frame contains data from only one channel at a time and each terminal is scanned in time sequence. This method works well for low traffic densities but it introduces delays as the traffic density increases. (*b*) A frame can contain data from different channels. Each channel has one, or two, bytes which indicates its terminal address and the number of characters in the frame. This method gives a higher transmission efficiency than the first method for low traffic densities but it is not as efficient as method (*c*) for higher traffic densities. (*c*) With this method between two to six position bits are assigned to each channel. When a channel is not active the frame contains the overhead of these position bits, but when the channel is active the data bits replace the position bits. This method has a transmission efficiency that increases with increase in the traffic density. (*d*) Now, rather than assigning a one-byte, or a two-byte, address to every active channel, a two-bit, three-bit or four-bit address indicates the position of the channel relative to the address of the preceding active channel. This gives a transmission efficiency somewhere in between that given by the methods (*b*) and (*c*).

Data Compression

The transmission efficiency of a statistical multiplexer can be increased by the use of *data compression*. The Huffman encoding technique reduces the average length of the characters in a message by converting the seven-bit ASCII characters into characters whose length varies between three bits and seven bits. The most frequently used characters are changed into three-bit words and least often used characters remain as seven-bit words. Other characters have varying lengths depending upon the frequency of their occurrence. The use of data compression can give about 30% reduction in the average length of the characters.

Example 5.4

The high-speed link between two statistical multiplexers operates at 14.4 kbit/s. Determine the number of characters per second that are transmitted if data compression is (*a*) not used, (*b*) used.

Solution
    (*a*) Number of characters = 14400/8 = 1800.   (*Ans.*)
    (*b*) Number of characters = 1800 + 1800 × 0.3 = 2340.   (*Ans.*)

Synchronous Signals

Synchronous signal are also applied to statistical multiplexers even though little, if any, increase in the transmission efficiency is thereby gained. Multiplexing allows the mixing of different synchronous protocols, and/or the mixing of synchronous and non-synchronous signals. Two ways in which a STDM may be operated are as follows. (*a*) When a synchronous terminal becomes active a predefined part of the bearer circuit's capacity, equal to the speed of the terminal, is allotted to it. This gives the advantage that the STDM is then insensitive to the protocol employed by the terminal but it has two disadvantages: (i) the aggregate speeds of the active terminals cannot exceed the bit rate of the bearer circuit; and (ii) the SYN and PAD characters will usually be retained so that the transmission efficiency will not be as high as it might be. (*b*) The data from each of the active terminals is multiplexed, whether or not it is synchronous, and it is then transmitted within internodal frames. Delay-sensitive synchronous data is given priority and a larger part of the capacity of the bearer circuit. This technique allows the aggregate data rate of the active synchronous terminals to be greater than the bit rate of the bearer circuit, but it does have the disadvantage that it takes longer to form the frames thereby causing some transmission delay.

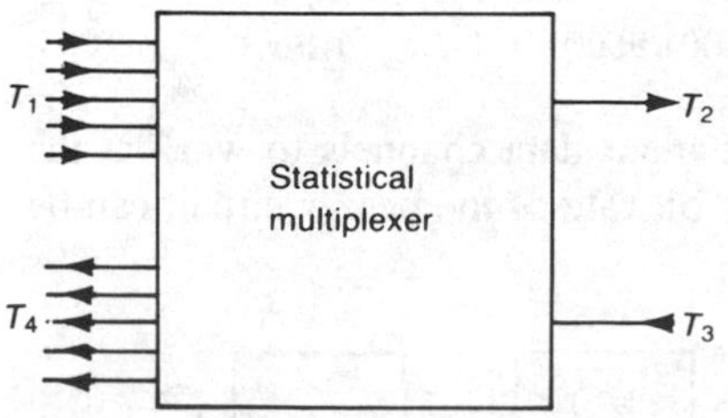

Throughput and Burst-mode Handling

The throughput of a STDM is the quantity of continuous data, in characters per second, which it is able to transmit. The burst-mode handling capability of an STDM is the instantaneous peak capability of the multiplexer. Figure 5.8 shows an STDM with five channels

T_1

T_2

Statistical
multiplexer

T_4

T_3

Fig. 5.8 Burst handling.

inputting data at a total rate of T_1 and being transmitted to the bearer circuit at a data rate of T_2. At the same time data, at the rate of T_3, is being received from the bearer circuit and transmitted into five channels at an aggregate rate of T_4. The throughput of the STDM is equal to $T_1 + T_2 + T_3 + T_4$. The burst-mode considers the channel inputs T_1 (or the channel outputs T_2) for a short period of time. A high burst-mode handling capability is required if many high-speed terminals with low traffic utilization are connected to the STDM, when the probability of all the terminals being active at the same time can be catered for. Typically, the burst-mode handling capability is about nine times the bit rate of the high-speed bearer circuit.

Example 5.5

A statistical multiplexer has 16×4.8 kbit/s channels and its bearer circuit operates at 9.6 kbit/s. Calculate its burst-mode handling capability.

Solution
Burst-mode handling capability $= (16 \times 4.8)/9.6 = 8.$ (*Ans.*)

A statistical multiplexer is able to provide service to a number of data channels whose aggregate bit rate is greater than the maximum bit rate that the bearer circuit is able to transmit. The ratio between the two bit rates is expressed by the *compaction* (or *compression*) *ratio* of the statistical multiplexer.

$$\text{Compaction ratio} = \left[\frac{\text{aggregate of input channel bit rates}}{\text{bit rate of bearer circuit}} \right]$$

$$(5.2)$$

Typically, the compaction ratio varies from about 2:1 for synchronous channels to about 7:1 for non-synchronous channels.

Example 5.6

Figure 5.9 shows a basic statistical multiplexer network. Calculate its compaction ratio.

Solution

Compaction ratio $= (8 \times 4800)/9600 = 4:1.$ (*Ans.*)

It is not necessary for all the input data channels to work at the same speed as one another. The bit rate of the bearer circuit can be

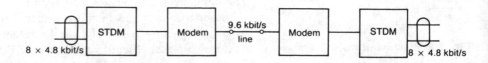

Fig. 5.9

split up, in various ways, and allotted to groups of input channels. If, for example, the bearer circuit's bit rate is 9.6 kbit/s and the input data channels consist of two at 2400 bit/s, three at 1200 bit/s, and six at 600 bit/s, then the bearer circuit's capacity could be divided: (i) 4.8 kbit/s for the 2400 bit/s channels, (ii) 2400 bit/s for the 1200 bit/s channels, and (iii) 2400 bit/s for the 600 bit/s channels.

A link set up between two statistical multiplexers will employ some form of error detection. As each frame is transmitted it is also stored in a buffer and is held there until a signal has been received from the distant multiplexer acknowledging the correct reception of the data. If an error has been detected, this fact is signalled back to the transmitting multiplexer and the corrupted frames are then retransmitted, again and again if necessary, until their correct reception has been acknowledged. This means that the terminals themselves need not include error-protection circuitry and they may therefore be relatively cheap non-synchronous terminals. Some statistical multiplexers are also able to function as switching multiplexers.

Concentrators

A concentrator is an equipment that employs the principle of contention. This means that a number of input channels are contending with one another for access to a smaller number of output channels on a demand basis. The concept is rather like that used with telephone switchboards and it is illustrated by Fig. 5.10. Contention takes advantage of the small probability of all the inputs to the concentrator being active at the same time. The concentrator polls each of its inputs in sequence to see whether they have data waiting to be transmitted. If a channel is active then it is switched to an output line and the distant modem is signalled that data is coming. When data is received from the distant end of the line the concentrator must determine the address of the destination terminal and route the data accordingly. For a concentrator the output bit rate is less than the aggregate input bit rate. Contention has two main advantages and two disadvantages.

(*a*) It gives a good performance with channels with long propagation delays.
(*b*) It is cost-effective for terminals with low traffic densities.
(*c*) It is inefficient for use with terminals that output high traffic.
(*d*) An individual terminal can easily occupy an output channel without sending any data.

Many concentrators also include one or more of a number of other functions such as the following.

(*a*) Data conversion. Some concentrators are able to change character codes, bit rates, and/or protocols and so enable otherwise incompatible terminals to be interfaced;
(*b*) Data compression.
(*c*) Store-and-forward facilities. Some concentrators are able to

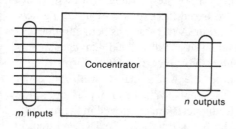

Fig. 5.10 Principle of a concentrator.

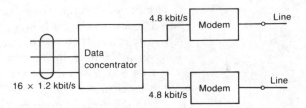

Fig. 5.11 Use of a concentrator.

store a whole message before it is transmitted and this can be helpful when some inputs are from non-buffered terminals. If more inputs channels than there are output channels send data simultaneously, some of the input data can be held in store until a free output channel becomes available and the data can be sent. Other concentrators are of the hold-and-forward type; here a terminal sends data to the concentrator only when commanded to do so and this makes sure that the instantaneous input data rate is never in excess of the maximum possible output data rate. Control of the input channels is achieved by polling.

(*d*) Message switching. Some concentrators are able to switch input data to any one of several destinations rather like a switching multiplexer.

(*e*) Processing. Some concentrators have the capability to perform some processing of data themselves which then reduces the amount of data that must be sent to the computer for processing.

(*f*) Error correction.

A concentrator that includes some of the above features in addition to contention is usually known as a *data concentrator*.

Figure 5.11 shows one way in which a concentrator may be used. Sixteen 1200 bit/s terminals are connected to the concentrator's input ports and there are two 4.8 kbit/s output lines. The output data is digital in nature and so it must be applied to a modem before it can be transmitted over the telephone network.

Data-Over-Voice Multiplexer

A data-over-voice multiplexer (DOV) allows the use of already existing telephone cables within an office building or a factory site to be also used for data and so avoid the expense of installing new cabling just for data communication. A DOV acts as both a frequency-division multiplexer and a modem; it converts the digital data signals from a terminal or a computer into voice-frequency form using a carrier frequency that positions the v.f. signal at well above the bandwidth occupied by speech signals. When full-duplex operation is required different carrier frequencies are employed in the modems at each end of the circuit. The block diagram of a DOV system is shown by Fig. 5.12.

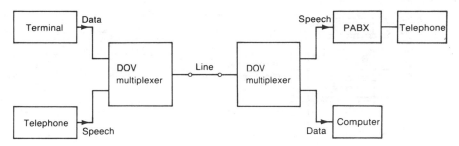

Fig. 5.12 Data-over-voice multiplexer system.

Digital PABX

All offices have long made use of a telephone switchboard or a PABX to allow telephones in various offices to have access to one another, and to the public telephone network without the expense of giving each telephone a direct line. The older types of switchboards and PABXs were analogue in nature and were only able to switch telephone calls. The more recent digital PABXs are able to provide a number of other modern services, such as data switching, electronic mail, and local area networks as well as switch telephone calls.

The basic concept of a digital PABX is illustrated by Fig. 5.13. The number q of lines from the PABX to the PSTN is much less than the number n of telephones since the telephones will not, on average, be making, or receiving, an outside call for most of the time. The digital PABX allows both the telephone extensions and the data terminals to *contend* for a PSTN line whenever one is wanted. Any of the n telephone lines can be connected to any other telephone line or to any one of the q PSTN lines. Any of the p data terminals can be connected to any of the m computer ports or to any of the q PSTN lines.

The digital PABX converts all speech signals into 64 kbit/s data streams that can then be switched in the same way as the actual digital data signals. The PABX provides a 64 kbit/s connection between any two of its $(m + n + p + q)$ ports. Each of these connections is able to carry either speech or data signals, or both speech and data signals simultaneously. The input analogue speech signals are converted into digital form by an IC known as a *codec*, which essentially is an

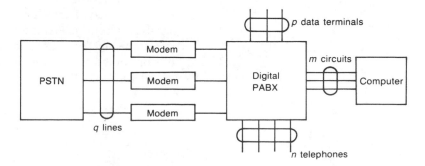

Fig. 5.13 Use of a digital PABX.

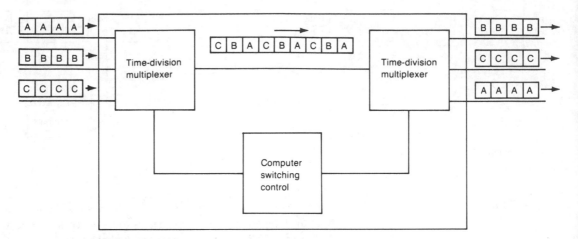

Fig. 5.14 Basic operation of a digital PABX.

analogue-to-digital converter in one direction and a digital-to-analogue converter in the other direction. The operation of a digital PABX is rather like that of a time-division multiplexer having 64 kbit/s inputs and a 2.048 Mbit/s common highway, except that computer control is employed to route the digital signals arriving at each input port to the wanted output port. The basic operation of a digital PABX is shown by Fig. 5.14.

Some terminals, such as workstations, are also provided with telephone facilities and the user of such a terminal will be able to speak over the telephone and, at the same time, have the terminals in communication with the computer. The switching of the circuits within the PABX is under the control of an internal computer which usually has some spare capacity. This spare capacity can be employed to carry out data processing for some terminals so that computer access may not be necessary; alternatively, the processing power may be used for such services as electronic mail. The digital PABX is transparent and is able to switch any form of data regardless of its format or of the protocol employed.

One advantage of the digital PABX is that it can replace a conventional PABX and use the existing in-house telephone cabling. This feature can provide considerable savings over the provision of a local area network which would require the building to be re-cabled with either coaxial cable or fibre optic cable. At present any data signals sent from the PABX to a distant destination via the PSTN may need to be converted to analogue form by a modem before transmission. In the near future, however, when the Integrated Services Digital Network (ISDN) is installed and working this will not be necessary.

6 Networks

The telephone lines, or other transmission media, employed for data transmission are very expensive to use. Consequently, various techniques are employed to enable the maximum possible data to be transmitted over the least possible number of lines. A compromise must be made between line costs, transmission speeds, and the requirements of each circuit, and, as a result, there are a number of different network configurations and techniques in use. The main alternatives to be considered in the design of a network are: dial-up connections via the PSTN, leased private analogue point-to-point circuits, Kilostream and Megastream, and packet switching. The first two of these alternatives use voice-frequency analogue signals and they require the use of a modem at each end of each circuit. The remaining three alternatives all use digital signals and so they do not employ modems. In many cases, other than the smallest networks, a mixed-medium solution to a network problem often turns out to be the most cost-effective.

The public switched telephone network (PSTN) has been designed for the transmission of analogue speech signals and it is unable to transmit digital data signals. It is therefore necessary to employ a modem at the transmitting end of a data circuit to convert the digital signals into analogue form, and to employ another modem at the receiving end of the circuit to convert the received signals back to their original digital form. The time taken to establish a connection via the PSTN is usually at least 15 s and is very often considerably longer, also calls are sometimes mis-routed. The transmission characteristics of the telephone lines making up the connection may be rather poor and there may be wide variations in the noise and interference that is picked up, and the distortion introduced, even for connections set up between the same two locations at different times. This is because calls may be routed over different paths in the telephone network, particularly at times of heavy traffic. This noise may cause errors in the received data and it will most certainly restrict the maximum speed of transmission unless the most expensive modems with error correction and echo-cancellation facilities are employed. Because of its method of charging for calls it will prove expensive to use the PSTN to transmit large quantities of data. A PSTN connection only provides a two-wire presented circuit; if a four-wire presented circuit is wanted then two PSTN circuits must be dialled up.

Some of the problems associated with the use of the PSTN may be overcome by the use of a non-switched point-to-point leased private analogue circuit. It will still be necessary to employ a modem at each end of the circuit but the line can now be conditioned to allow higher-speed transmissions using simpler, and hence cheaper, modems. Furthermore, the call set-up time will be very short and line costs are fixed regardless of the amount of data that is transmitted. Various line-sharing techniques, such as clustering, multi-dropping, and multiplexing can be employed to still further reduce line costs.

Network Topology

The way in which a private data network interconnects a number of terminals and computers is known as its *topology*. In modern data networks several different topologies are commonly employed.

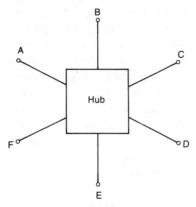

Fig. 6.1 Star network.

Star Network

A star network is shown by Fig. 6.1. A number of terminals are connected to a central point, or hub, by separate circuits and all inter-terminal connections must be routed via the hub. The most common application of the star network is the connection of both local and remote terminals to a host computer or to a digital PABX. The operation of the network is completely dependent upon the correct operation of the hub and if it should develop a fault the whole network will be affected. Should the data traffic be heavy there may be considerable delays because of congestion in the hub.

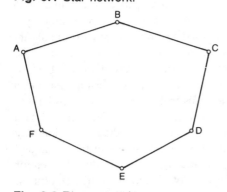

Fig. 6.2 Ring network.

Ring Network

In a ring network, see Fig. 6.2, each switching point or *node* is connected to its two adjacent nodes. All the messages to be transmitted are addressed before they are transmitted into the ring. They are then circulated around the ring until the destination terminal recognizes its address and accepts the message. A single break anywhere in the ring will not put it out of operation if signals can still be transmitted in the opposite direction. The ring topology is rarely used for any data network other than a *local area network*.

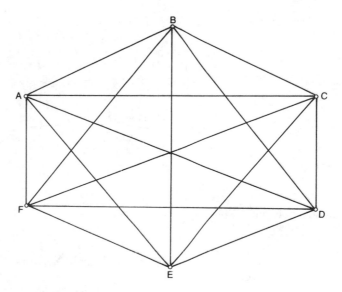

Fig. 6.3 Mesh network.

Mesh Network

A mesh network has a number of fully interconnected nodes, see Fig. 6.3, and this allows it to provide a multiplicity of possible paths for any required inter-terminal connection. The multiple message paths reduce the effects of any link and/or node failures and/or congestion which may occur since any desired connection can always be rerouted. The main disadvantages of the mesh network are the expense of interconnecting all the nodes and possible transmission delays; because of the cost many mesh networks are not fully interconnected.

Bus Network

Figure 6.4 shows that a bus network has a number of terminals connected to a main line. This topology is used in local area networks and for multi-point networks over relatively short distances. It possesses the advantage that extra terminals can easily be added to the network without any need to reconfigure the network.

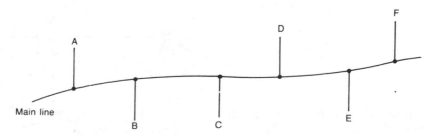

Main line

Fig. 6.4 Bus network.

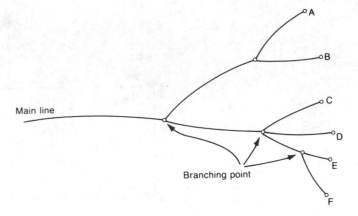

Fig. 6.5 Cluster network.

Tree, or Cluster Network

A tree, or cluster, network is one in which a computer is connected by a four-wire line to a branching point, known as a cluster controller, that is situated in a telephone exchange. The cluster controller splits the line into two, or more, branches and each of these, in turn, is split into two, or more, branches by other cluster controllers and so on up to a maximum of twelve *spurs*. The topology of a cluster network is illustrated by Fig. 6.5 and it is often employed to connect remote terminals to a computer. One modem is necessary at the computer end of the main line and one modem at the terminal end of each spur. A cluster network is economical in both line and modem costs.

Very often a data network employs some combination of two, or more, of the topologies mentioned; for example, Fig. 6.6 shows a combination of the star and the cluster networks.

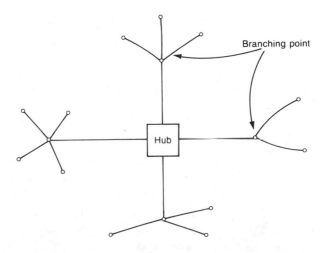

Fig. 6.6 Combined star and cluster network.

Switching

Larger data networks generally employ some kind of switching so that the network can be shared by many users. There are three different types of switching that are employed.

(a) In *circuit switching* an end-to-end link must first be set up and then the message is transmitted; when the message has been received the calling terminal releases the circuit and the connection is disconnected.

(b) In *message switching* a message is sent into the network with the address of its destination added and it is routed through the network to its destination as soon as possible.

(c) In *packet switching* messages are transmitted into the network and are then broken up into a number of separate packets. Each packet is labelled with the address of its destination and it is then transmitted through the network as a discrete entity. At the output of the system the message is re-assembled from the received packets and is then transmitted to the destination terminal.

Both message and packet switching are considered in Chapter 8.

Four digital services (under the group name X-Stream) that do not require the use of modems and give a better performance than circuits routed via the analogue network are offered by BT. The four services are:

(a) Kilostream which operates at 64 kbit/s over digital links that provide point-to-point links between the larger telephone exchanges;

(b) Megastream which operates at 2 Mbit/s, 8 Mbit/s, 34 Mbit/s and 140 Mbit/s to provide links that are used, for example, to interconnect two digital PABXs or for high-speed applications such as teleconferencing, slow-scan television and remote CAD/CAM;

(c) Satstream which offers high-speed digital circuits via communication satellites to destinations in other countries; and

(d) Switchstream which is a switched service that uses packet switching.

Point-to-point Analogue Networks

In a point-to-point analogue network each terminal is connected to the computer by a direct circuit that is exclusive to that terminal. A dial-up connection via the PSTN is, of course, a point-to-point circuit. A leased point-to-point circuit offers the advantage that the computer and the terminals have immediate access to one another, and there are many applications where this feature is of paramount importance.

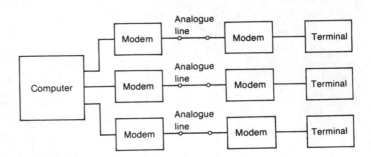

Fig. 6.7 Point-to-point data circuit.

The computer software is simplified because any data transmitted by the computer from a particular port can only go to the one terminal; conversely, all the data arriving at a specified computer port will be known by the computer to have originated from a particular terminal. The speed of response is fast because the maximum speed and capacity of the circuit is used for only the one terminal. The arrangement of a point-to-point circuit that uses analogue lines is shown by Fig. 6.7. Each computer port is connected to a modem that converts the data signals into voice-frequency form before they are transmitted over the line. At the distant end of each circuit the received signals are changed back into their original digital form by the modem and are then passed on to the terminal.

For most computer-to-terminal links point-to-point operation is extremely inefficient and hence costly, because most of the time a circuit will not be in use since neither the computer nor the terminal will have data to send.

Polled Networks

The utilization of a telephone line can be increased by connecting several terminals to the one line. Two methods of doing this are shown in Figs 6.8 and 6.9 and are known as *clustering* and *multi-dropping*, respectively. Clustering means that several terminals are connected to the line at the same point by means of a cluster controller. All the clustered terminals must be located within a few kilometres of the junction. Multi-dropping means that a number of terminals are connected to a line at different points along its length. The number of 'drops' is limited to seven. Very often a combination of both

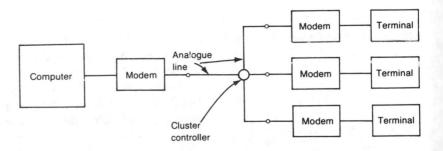

Fig. 6.8 Cluster network.

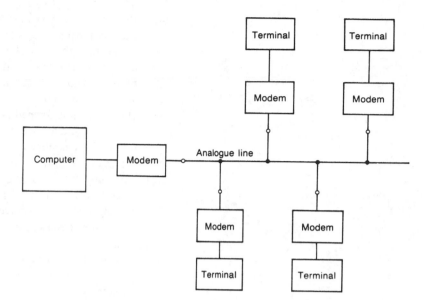

Fig. 6.9 Multi-dropping network.

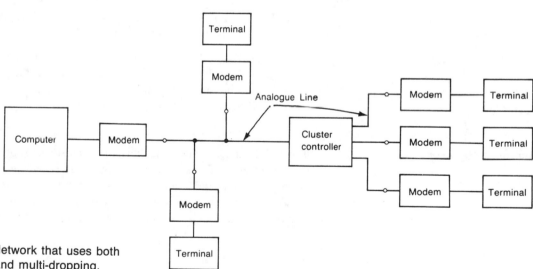

Fig. 6.10 Network that uses both clustering and multi-dropping.

clustering and multi-dropping is used and Fig. 6.10 gives an example of this technique. In all three circuits the block marked as terminal may represent a single terminal or it may represent several lower-speed terminals combined together in some way. This might involve, for example, the use of a multi-port modem or a modem sharing unit.

Computer software is required to enable a computer to select each terminal in turn so that it can send and/or receive data to/from the computer. This is known as *polling*. There are four main kinds of polling in use.

(a) Roll call. With this system the computer sequentially invites each terminal to access the line. If a terminal has data to send it transmits the first block of that data; if it has nothing to transmit it returns a 'no-data' signal to the computer and this returns control of the line to the computer and the poll sequence continues.

(b) Hub. A loop is formed between the computer and all of the terminals, and a single poll is sent to the first terminal. This terminal transmits its data (if any) to the computer and then it passes the poll on to the next terminal.

(c) Selecting. While polling is in progress the computer is able to access the line to verify that a terminal, not currently polled, is ready to receive data; data can then be sent to that terminal out of sequence before the polling is allowed to continue.

(d) Broadcasting. The computer transmits a poll message to the network that is received by all the terminals. The poll includes the address of the called terminal and only that terminal responds to the poll.

The use of polling gives the advantage that each terminal has the full use of both the line and the computer when it is on-line. Polling has, however, the following disadvantages.

(a) There may be some delay in the response from the computer when several terminals require service at the same time.

(b) Polling software is necessary.

(c) A line fault will affect several terminals.

Use of Multiplexers

The use of a multiplexer, time division or statistical, allows a number of low-speed terminals to share a single high-speed line with little, or no, extra delay being introduced, and without the need for special computer software. The basic arrangement of a multiplexed circuit is shown by Fig. 6.11. Five computer ports are multiplexed and the

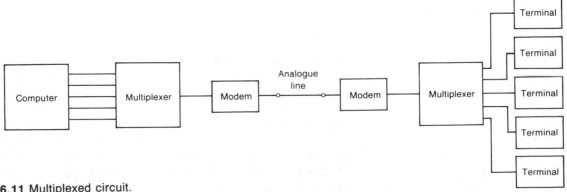

Fig. 6.11 Multiplexed circuit.

aggregate signal is applied to a modem and changed into a voice-frequency signal before it is transmitted over the line. At the distant end of the line the received signal is converted back to digital form by the modem and then it is demultiplexed to obtain the original five separate data signals. Clearly, it is necessary for the operation of the receiving multiplexer to be synchronized with that of the transmitting multiplexer to ensure that the received data streams are routed to the correct terminals. A slightly more complex multiplexed circuit is shown by Fig. 6.12.

Some multiplexers are able to handle a mixture of both synchronous and non-synchronous signals and some types are provided with the ability also to act as data switches. The use of switching multiplexers allows the provision of a multi-node network, and Fig. 6.13 gives a relatively simple example. The network includes five switching multiplexers, labelled as A, B, C, D and E. The central node A receives five high-speed data streams from the computer and it is able to switch any stream to any one of the distant nodes. In turn, each of the distant nodes can switch any of its inputs channels to any one of a number of low-speed terminals. The traffic to, and from, a terminal tends to be sporadic in its nature, and so the loading on the computer is much more evenly spread if there are many more terminals than there are computer ports and each terminal has to contend with the other terminals for access to a port. In some networks a terminal may be able to gain access to a particular computer port. For many organizations the reliability of their private data network is very important and the business might grind to a halt if the network should fail. Many modern networks therefore are no longer point-to-point, as in Fig. 6.13, but instead employ some form of mesh network.

Switched Networks

A large mesh data network will probably use one, or more, data switches of some kind, including switching multiplexers. A multi-mode network that has one, or more, computers, and a large number of terminals of various types, located at different locations is often known as a *wide-area network* or WAN. The switching technique employed, which is very similar to that used in telephone exchange networks, is known as *circuit switching*. To establish a link between two terminals a call request signal must be transmitted by the calling terminal. There will be some delay at each switching point before the request reaches its destination. If the called terminal is free and able to receive data an acknowledgement signal will be returned and then the transfer of data can begin. A typical WAN is shown in Fig. 6.14. The network has two data switches and three switching multiplexers to make it possible for any terminal to be connected to any port on either of the two computers. Considerable flexibility is provided in the possible routings between any two locations in the network, which is of great assistance in the event of traffic congestion on some links and/or faulty lines. If one of the main lines should fail

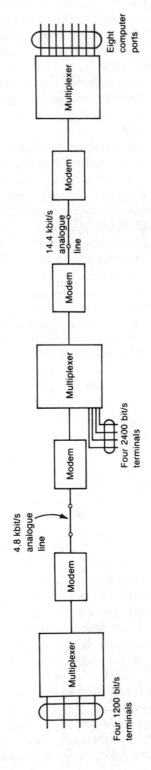

Fig. 6.12 Multiplexed circuit.

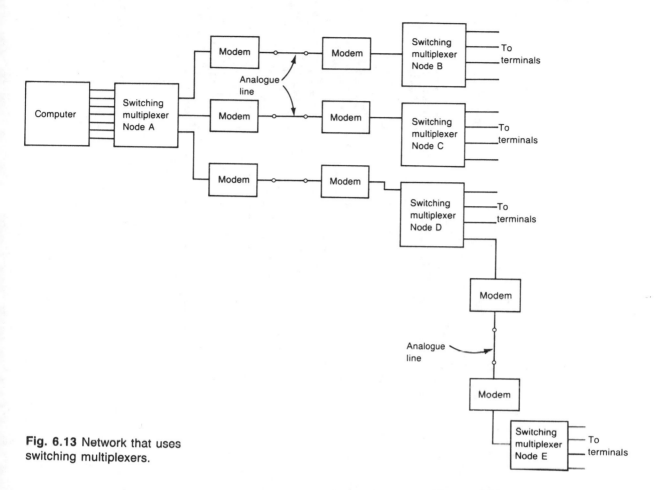

Fig. 6.13 Network that uses switching multiplexers.

whilst it is in use a terminal can be switched on to some other route and still be given access to a required computer port.

The advantages of circuit switching are transparency and very low transmission delays, and it is attractive for networks that carry a more, or less, constant traffic and/or bulk data. The circuit switching of data possesses the following disadvantages.

(a) Individual circuits are established and maintained that have a permanently allocated transmission capacity. This gives very inefficient usage of the line capacity for burst-like data applications since sometimes no data is being transferred over the circuit.

(b) Error control is not provided.

(c) Speed conversion is not provided, which means that the two ends of any connection set up in the network must be able to operate at the same speed as one another.

(d) Messages are not stored within the system.

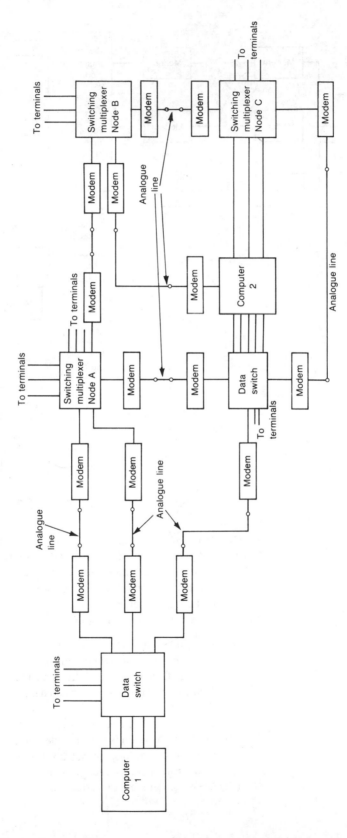

Fig. 6.14 A wide-area network (or WAN).

Data Switches

The choice of the type of digital switch to be used in a data network depends upon whether the data to be switched is of synchronous or non-synchronous form. Switching a non-synchronous data signal is relatively easy since only the transmitted data line, the received data line, and one or two control lines need to be switched. There are no direct timing relationships between these lines and so any delays that might be introduced are unimportant. On the other hand, a synchronous signal requires the switching of, at the very least, the transmitted data and received data lines, the transmit clock and receive clock lines, plus up to 12 control lines. The timing relationships between these various lines are critical and any differences in delay caused by switching, or jitter, cannot be tolerated. When switching a synchronous signal all the 16, or more, lines must be treated identically, no distinction being made between data, clock and control lines. A data switch must be centrally controlled, transparent, able to allow all inputs to be connected to all outputs, and reliable in its operation.

Non-synchronous

Typically, a non-synchronous data network contains a number of small computers, each of which supports a large number of terminals that have direct access to a port. Each of the terminals will appear transparent to the computer. Each computer can only run a small number of concurrent programs and support a small number of ports. This means that the terminals must contend with each other for access to a port and that each computer must be restricted to a fairly small number of tasks. The users of each terminal must then select the appropriate computer in the network to perform a particular task. Thus the main requirements for a non-synchronous data network are: (*a*) transparent terminals; (*b*) flexible users; (*c*) port contention; (*d*) short access time; and (*e*) computer selection. There are five main ways in which these requirements can be satisfied, i.e. switching multiplexers, matrix switches, data PABXs, voice/data PABXs and local area networks.

Switching Multiplexers

The switching nodes are distributed around the network and each node employs a switching multiplexer. The nodes are interconnected using normal communication circuits. The main disadvantage of this method of switching is that a failure that occurs in any part of the network may have repercussions on other parts of the network.

Matrix Switches

A matrix switch has a number of input circuits and a number of output circuits connected to a matrix which consists of a number of lines that are arranged mutually at right angles in the horizontal and vertical planes. The intersections between the horizontal and the vertical lines are known as *crosspoints* and they can be connected together by electronic switches. Any one of the input circuits can be connected to any one of the output circuits by the closing of the appropriate electronic switch. The number of crosspoints is equal to the product of the number of input lines and the number of output lines. It is not necessary for the number of inputs and outputs to be equal to one another. A matrix switch that is employed to switch data signals must be: (*a*) able to operate at the highest bit rate used by any of its input and output circuits; (*b*) able to provide non-blocked connections between inputs and outputs, i.e. all ports can be interconnected in pairs at the same time; (*c*) able to monitor all connections; and (*d*) be transparent to data, clock and control signals so that there is no need for the data speed and/or the protocols to be specified.

Data PABX

A data PABX is a transparent multiplexer-based switching system that uses a high-speed bus with a limited number of circuits to link together its input and output ports. Data signals arriving at an input port are time-division multiplexed with other input data signals and then the composite signal thus produced is passed over a high-speed bus to the output side of the switch. Here the composite data signal is demultiplexed to obtain the individual data signals and these are directed on to their particular wanted output lines. The high-speed bus is able to transmit a mixture of data, clock and control signals.

Digital Voice/data PABX

When speech signals are applied to a digital voice/data PABX they are first converted to 64 kbit/s digital signals, using a technique known as *pulse-code modulation* (PCM), and then they are time-division multiplexed with the data signals before being applied to a high-speed bus. The multiplexed signals are then routed through the PABX over its high-speed bus to their destination output ports. The high-speed paths through the PABX can carry high-speed data signals and it is not efficient to use them to carry low-speed data.

A PABX is suitable for

- (*a*) The connection of dumb terminals to mainframe computers and to external networks,
- (*b*) Connecting terminals to LANs, and
- (*c*) The integration of voice and data communications.

A PABX is not suitable for use with systems that require a wide bandwidth or that do not require switched connections. The benefits that arise from the use of a PABX in a data network are:

- (*a*) No separate cabling of an office is necessary,

(*b*) Easy to re-configure; as staff in an office are moved around it is only necessary to re-allocate the ports on the PABX,

(*c*) Provides uniform access to external data networks,

(*d*) Since circuits can be used by both voice and data signals, use of the network is optimized.

Synchronous

A synchronous switched network is mainly employed for applications where batch and/or long-duration interactive enquiry/response data transfers are common. Many synchronous terminals can be connected to each port of a computer so that each terminal must have its own unique address. The computer must be programmed to know which addresses are connected to which ports in order that it will be able to direct data to the correct port. The three main factors to be considered in the design of a synchronous switched network are the following.

(*a*) Any changes in the configuration of the network must be easy and quick to implement.

(*b*) Synchronous networks are always complex and the detection and location of problems can be difficult. Therefore a test and diagnostic centre is usually provided.

(*c*) Usually, synchronous terminals are required to be on-line at all times and this means that standby equipment which can be switched into service if and when a fault occurs, is necessary.

Front-end Processor

Mainframe computers are designed to support a very large number of dumb terminals, many of which may be located some considerable distance from the computer. The majority of mainframe computers employ a *front-end processor* (FEP) to carry out most of the communications protocols and switching functions. This practice ensures that the computing power of the mainframe is kept available for processing all the data inputted to it. An FEP usually includes a multiplexer and it often communicates with a *cluster controller* which runs the communications protocols. The terminals need only then deal with such matters as keyboard entries and screen displays.

Nowadays many terminals are actually PCs running emulation software, and are hence intelligent. These must be provided with a communications protocol that allows file transfers.

The control of a communication network can always be given to the computer itself but the necessary software reduces the information-processing capability of the computer. Very often, therefore, the control of the communications of a network is given to an equipment that used to be called a front-end processor but today, when more facilities have frequently been added, is more often known as a

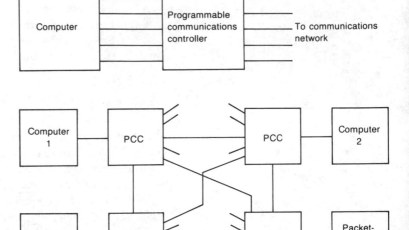

Fig. 6.15 Use of a programmable communications controller.

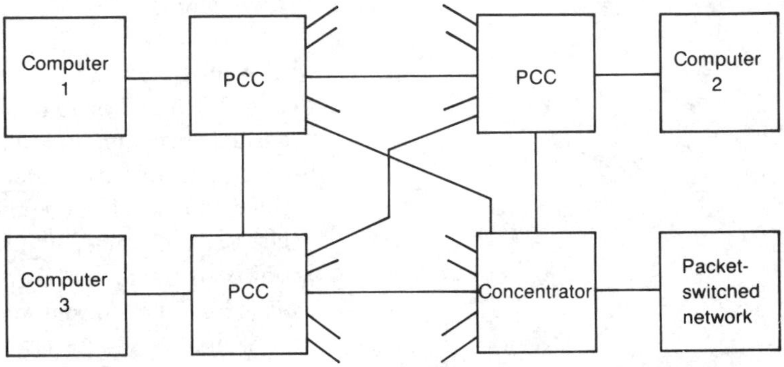

Fig. 6.16 Large network uses several PCCs.

programmable communications controller (PCC). A PCC acts as the logical centre of a network and it has complete control over the transfer of data from the computer to a terminal or vice versa. This arrangement allows the computer to apply all of its processing capacity to the processing of its input data. The PCC is connected between the computer and the network, as in Fig. 6.15. A large network that contains several computers will employ several PCCs and an example of this is given by Fig. 6.16. When several PCCs are needed it may be more efficient to employ one larger controller instead; this would be situated somewhere near to the geographical centre of the network and would be known as a *communications network controller* (CNC).

Protocol Conversion

Synchronous terminals can be interconnected using a polled network and/or by the use of multiplexers, but these techniques are not entirely satisfactory when the network contains a mixture of synchronous and non-synchronous terminals. Furthermore, these two techniques do not usually permit a terminal to access more than one type of computer. For any large network, in which there may be several types of computer, such limitations are undesirable. The difficulty can be overcome by the use of *protocol emulation* ('spoofing') or *protocol conversion*. A protocol emulator is software that is run in a microprocessor situated at the channel interface and which is able to communicate with both the terminal and the computer by using the appropriate protocols. This 'spoofing' makes the network appear to the terminal as its own kind of computer, and, at the same time, makes the network appear to the computer as its own type of terminal. An example of the technique is shown by Fig. 6.17.

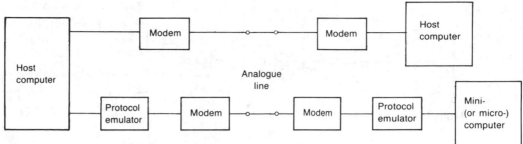

Fig. 6.17 Use of a protocol emulator.

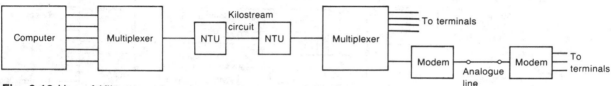

Fig. 6.18 Use of Kilostream in a data network.

Networks using Digital Circuits

Digital circuits, known as Kilostream and Megastream by BT, provide high-speed point-to-point circuits which can be used as integral parts of private data networks. A simple network using Kilostream is shown by Fig. 6.18. The computer's output ports are multiplexed together to form a high-speed data stream and then this is transmitted, via a network terminating unit (NTU), on to the Kilostream circuit. The network terminating unit is a short-distance synchronous line driver which converts the PCM signals used on the Kilostream circuit line into the standard ITU-T V24 (EIA 232) voltages used by the computer, or terminal, interface. The NTU transmits digital signals over a four-wire local line at either 12.8 kbit/s or 64 kbit/s using WAL2 encoding. The interface between the NTU and the customer's equipment follows the ITU-T X21 or X21 bis, recommendations and operates at 2.4, 4.8, 9.6, 48 or 64 kbit/s. A similar service is available from Mercury which offers bit rates of up to 256 kbits/s.

At the other end of the circuit the signal is demultiplexed and the individual channels are directed to their correct destination terminals. The terminals may be either synchronous or non-synchronous in their operation, or a mixture of both. The use of multiplexers together with the high traffic-carrying capacity of the Kilostream services allows cost-effective private networks to be assembled.

Figure 6.19 shows one method by which a computer can be connected to a large number of remote VDUs; this is the kind of arrangement that is used, for example, by many travel agents to access package holiday firms and air lines to check hotel vacancies, flight times, etc. A digital sharing unit (DSU) is a device that allows a number of data terminals to share a single Kilostream circuit, usually on a polled basis.

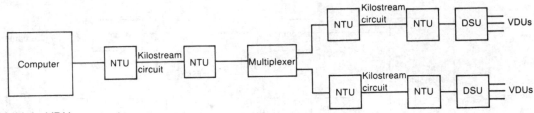

Fig. 6.19 Multiple VDU access to a computer.

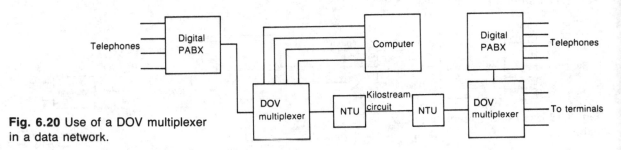

Fig. 6.20 Use of a DOV multiplexer in a data network.

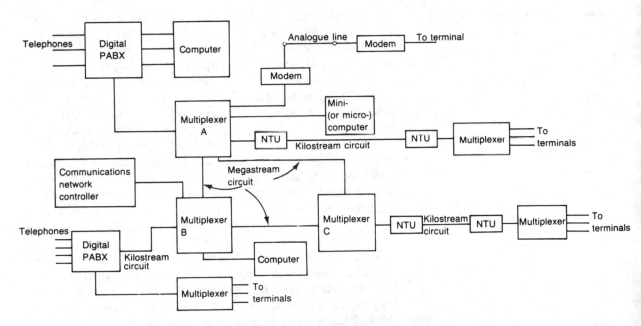

Fig. 6.21 Use of both Kilostream and Megastream in a WAN.

Data-over-voice (DOV) composite signals can also be transmitted easily over a Kilostream circuit and one possible circuit arrangement is shown by Fig. 6.20. The aggregate bit rates from both the speech PABX and the computer are each 32 kbit/s and when these are combined the maximum bit rate of 64 kbit/s is obtained.

The high bit rate provided by a Megastream circuit can be used to interconnect two data PABXs, or the capacity may be multiplexed to provide a large number of low-speed channels. Figure 6.21 shows a mesh network that uses both Kilostream and Megastream circuits to give a large number of data terminals access to two host computers and to a minicomputer; telephonic communication is also provided between two of the sites. The links from a multiplexer to data terminals that are not shown may be local, perhaps using a line driver or some other non-modem equipment, or more distant via two modems and a telephone line. The latter might, in turn, use a concentrator, a multiplexer, or even a multi-port modem, to further increase utilization of the line. Each of the three Megastream multiplexers in the network, labelled A, B and C, acts as a switching node and is able to handle up to 10 links with a maximum aggregate bit rate of 9.6 Mbit/s. Efficient control of the communications within the network is exercised by the network communications controller which is situated at node B. From this point the network manager is able to diagnose faults, collect traffic statistics, and, if necessary, reconfigure the network to cope with traffic congestion and/or faults.

An example of a complete data network is given by Fig. 6.22. It can be seen to include examples of many kinds of equipment including multiplexers, modems, data exchanges, and protocol emulators as well as both analogue and digital lines. The block labelled PAD, which stands for packet assembly/disassembly device, enables terminals that do not have an interface suitable for direct connection to a packet switched network to access such a network. As well as converting the terminal's usual data flow to and from the packet switching network the PAD handles all aspects of call set-up and addressing.

Mobile Data Networks

Most mobile telephones in the UK are connected to one of two cellphone networks, known as Cellnet and Vodaphone. It is possible to use a mobile telephone for data communications although it may sometimes be necessary to (a) be stationary (or at least not be in a moving car), and (b) be at a point where reception is good. This is because the radio signal path from mobile telephone to base station is usually much worse than the standard PSTN connection and its quality may fluctuate with location of the mobile.

The modems used for cellular data communications are specially designed to interface with a mobile telephone and employ a specialist error-correcting protocol. It is normally necessary to employ a package that consists of a mobile telephone, a modem and a notebook PC, although some notebook PCs have a built-in modem.

To achieve a reasonable BER it is necessary to employ complex error detecting and correcting protocols; one such protocol is the *Cellular Data Link Control* (CDLC). This generally results in a rather low bit rate of 4800 or 2400 bits/s, although if reception is poor this may be still further reduced. A low bit rate means that a dialled-up mobile circuit must be held for a relatively long time and, since mobile calls are expensive, mobile data costs may be high. Because of this the most common applications of mobile data systems are FAX and

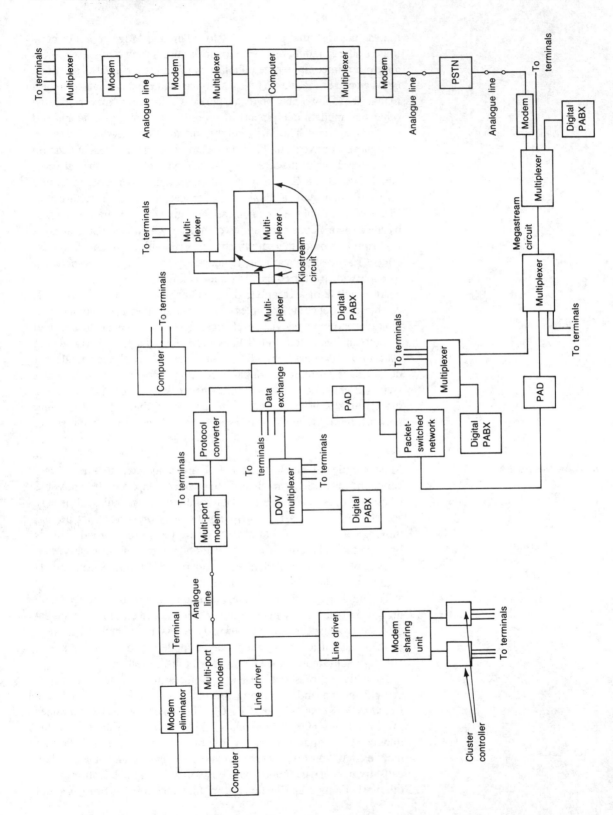

Fig. 6.22 Data network.

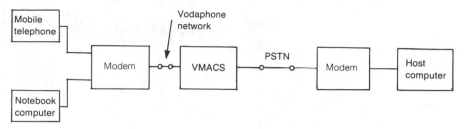

Fig. 6.23 Mobile data circuit.

Electronic Mail (p 129). A CDLC modem incorporates auto-dial and auto-answer and conforms with the ITU-T V25 bis dialling commands. The interface to the mobile terminal follows the ITU V24/EIA 232E recommendations.

The basic block diagram of a Vodata mobile data circuit is shown by Fig. 6.23. The mobile telephone and the notebook computer are connected via the CDLC modem to the Vodaphone network. The user 'dials' the telephone number of the host computer and is connected via the Vodaphone network to a *Vodaphone Mobile Access Conversion Service* (VMACS) equipment. The VMACS allows a user to initiate a data call between a mobile telephone and a host computer using the PSTN, and it also includes a testing procedure which is able to test a set-up connection before it is used to transfer data. When the number of the host computer is received by the VMACS it connects the caller to the wanted host computer, either via the PSTN or over a direct leased circuit.

The CDLC protocol is used over the radio section of a connection and the V42 protocol over the PSTN.

Many of the problems associated with mobile data communications are expected to be overcome once the GSM digital mobile telephone service has been established.

If voice communication is not needed as well as data communication a digital radio data system can be used that employs a technique known as *packet switching* (see Chapter 11).

7 Value-added Networks

Telephone and data networks, both public and private, are provided to give speech or data communication between two points. These two points may be at fixed locations or, if the network includes one, or more, switching points, any user may be able to access any other user. Since the de-regulation of the telecommunication services in the UK, and some other countries, an operator has been able to lease lines and equipment, such as modems, from the telephone authority, add some kind of extra service, or *value*, and resell the capacity. The network then offered together with the associated service is known as a *value-added network* (or VAN). A VAN service, or value-added service (VAS) is a service that adds value to the basic telecommunication network in order to provide a more cost-effective service.

A data network is said to have added value if it offers the customer something more than just the transmission of signals between two terminals connected to that network. The supplier of a VAN leases communication facilities from the telecommunications authority (BT and Mercury in the UK), puts them together to form a network and then adds *value*. It is this added value that distinguishes a VAN from a normal network, i.e. a LAN or a WAN. The information sent over a VAN must provide some kind of service such as aircraft seat or holiday booking, home banking or home shopping.

The user of a VAN is offered the use of an established data network as and when it is required, for only a small fraction of the cost of setting up, and operating, his own private network. Most VAN operators employ a combination of leased private lines, spare capacity on public or other private networks, and dedicated network controllers to provide the basic network that gives access to one, or more, computers. The VAN is normally accessed using either a computer or a viewdata terminal, and a modem. The main characteristics of a VAN are as follows.

(a) It transmits information over a public network such as the PSTN and/or a private data network.

(b) The information transmitted, which may be data, graphical, textual, video or voice, must be enhanced in some way so that the user is provided with a service.

(c) The control of a VAN is vested in a host computer.

(d) Some VANs allow the user to participate in the real-time sending and receiving of information, i.e. it is interactive.

Amongst the main types of VAN are electronic mail, protocol gateways, managed data services, electronic data interchange, on-line databases and viewdata, but there are also many others.

Electronic Mail

Electronic mail is the telecommunications equivalent of the ordinary postal service which delivers letters to houses and offices. The postal service is employed for the delivery of messages that do not need to be delivered within a very short time of their dispatch. Much business correspondence consists of orders, invoices, and receipts which are prepared using a computer at the originating office and then dealt with, at the receiving office, with the aid of another computer. Clearly, both time and money could be saved if the data could be transferred directly, via a data network, from one computer to the other. Any necessary monitoring, checking, or authorization could easily be carried out using a VDU and a keyboard. Electronic mail also has advantages for the passing of messages that could otherwise be telephoned. Very often a lot of time and effort can be spent trying to get through to a busy telephone, or FAX, number and even when the attempt has been successful the person wanted may not be in the office. The difficulty is that for a telephone conversation to take place both the parties must be in their offices, and at their telephones, at the same time. The use of electronic mail overcomes this problem. A user can send a message into the network and then leave it to the system to deliver the message to the destination address as soon as possible, (usually just a few minutes). If the recipient checks his mailbox frequently he will find the message waiting there for him. Thus, electronic mail is a system which provides its users with a cheap, rapid and convenient method of passing messages between, or within, offices, factories, etc.

When electronic mail first became feasible several different systems were employed. Each system worked perfectly well within its own network but it was incompatible with the other systems. To overcome this difficulty the ITU-T have introduced the X400 electronic mail recommendations which provide three protocols to enable messages, using either ASCII or binary code, to be transferred between networks. It also allows an electronic mail system to interconnect with the Telex, Teletex and FAX systems. The X400 standard defines a naming and addressing structure, a *user* agent (UA) and a *message transfer agent*

(MTA), which act, respectively, just like a US mailbox and sorting office. The three protocols are as follows.

(a) P_1 describes the envelope standard for the transmission of messages between two MTAs.
(b) P_2 defines the message format for both the header, or address, and the body, or content, of the message between two UAs.
(c) P_3 defines the envelope standard for communication between a UA and a MTA.

The architecture of an X400 message handling system (MHS) is shown by Fig. 7.1 and it can be seen to include both user agents and message transfer agents. Each user interacts, via a data line, with a user agent, or if Telex or Teletex is used with either a TLXUA or a TTXUA. The interaction prepares the message and then submits it to the system for routing to its destination(s). The UA submits each message to the associated MTA for transmission to the MTA nearest to the destination UA. The destination UA may be linked to the same MTA as the originating UA, or it may be anywhere else in the

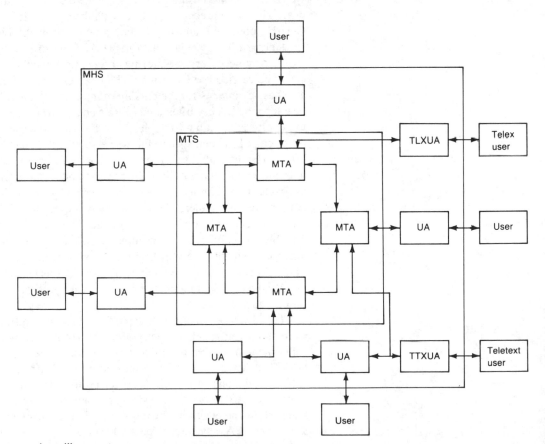

Fig. 7.1 Message handling system.

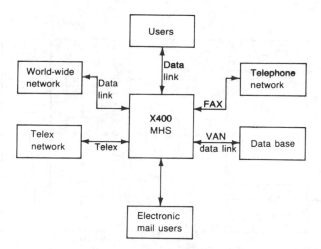

Fig. 7.2 Access to a message handling system.

network. In the second case the MTA must determine a route through the network and then relay the message on to the destination MTA. Here, the message is passed onto the destination UA and thence to the mailbox of the addressed user. In some cases the message might be relayed via several MTAs before it reaches its destination. The X400 specification is concerned with the passage of messages between a UA and its MTA, and between MTAs, but not with the interaction between a user and his local UA. A message may be stored in a queue at various points in the MHS if a route is congested and then forwarded when the wanted route becomes available. Most messages arrive at their destination mailbox within a few minutes but there is always some delay so that real-time interactive 'conversations' are not possible. The various ways in which a X400 MHS can be accessed are shown by Fig. 7.2.

A company may have several of these X400 electronic mail systems and may wish to connect them together to form a network that is known as a *private management domain* (PRMD). The company can have as many MTAs in a single PRMD, and also as many PRMDs in its network as it wishes. Also in existence is a public network of MTAs and this is called the *administrative management domain* (ADMD). A message that is to be passed from one PRMD to another may be routed either directly via the PSTN or via the ADMD as shown by Fig. 7.3.

An X400 electronic mail system provides the following features.

(*a*) Mail can be sent to and received from a single address, or sent to all the addresses that are on a mailing list.

(*b*) If a route within a MHS is busy the message can be stored at an intermediate point and forwarded on at a later time when the route becomes available.

(*c*) A message can be held within the network and not delivered to its destination address until a specified time.

(*d*) The sender of a message can be notified of its delivery, and/or

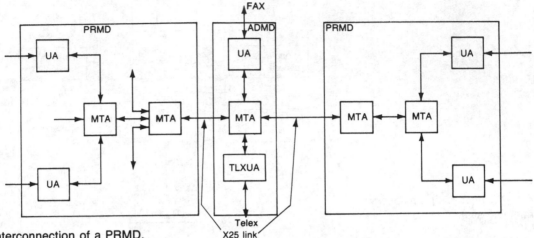

Fig. 7.3 Interconnection of a PRMD.

its non-delivery, to the destination mailbox, and/or the collection of the message from the mailbox by the recipient.

(*e*) When a user is going to be absent from his address he can inform the system that all his mail should be redirected to another specified mailbox.

(*f*) The sender of a message can request a reply by a certain date and/or time.

In the UK the electronic mail system provided by BT was Telecom Gold. The ADMD is known as Gold 400. The Telecom Gold service has now been superseded by *BT Mailbox*, although Telecom Gold and its existing customers are still supported. BT Mailbox uses BT's X400 message handling system to allow users to send and receive messages to other X400 compatible mailboxes around the world. Customers can also communicate with FAX and Telex systems world-wide, as well as to Dialcom systems that give access to on-line databases.

Other VANS

Viewdata

Viewdata, or videotext, is an interactive information-retrieval service which enables a page of data to be transmitted in one second. The UK service is provided by BT and is known as Prestel. The service offers the customer a range of computer-based information and communication services that are displayed upon the screen of a television receiver. A customer can gain access, via the PSTN, to the Prestel service at any time of day. Prestel consists of a network of computers that have been organized to store frames of information that are transmitted over a telephone line when requested by the customer. The customer selects the information frame wanted over a backward 75 bit/s channel. The information in the database is provided by a large number of sources, some of whom charge for the information given whilst others provide their information free.

Amongst the services that are offered are: airline, rail, and tour operators' information, seat booking and ticket issuing, stock exchange prices; up-to-date news, home banking; and software for home computers. Prestel can also be used to provide a two-way Telex service without the need to have an expensive Telex machine and line.

On-line Databases

An on-line database contains a large amount of information that can be accessed by a customer of the service. Databases vary in their content from the highly specialized to the fairly general and one example of the latter is BT's Prestel service

Protocol gateways

A protocol gateway makes it possible for different types of computer and/or terminal to communicate with one another, even though their protocols and their transmission speeds may differ. For example, some VANs use the gateways provided by BT's Prestel service to enable any user of a VAN to talk to any Prestel-compatible computer.

Managed Data Network Services

A managed data network service (MDNS) provides its users with such facilities as speed conversion, protocol conversion, packet assembly and disassembly, and data security. An MDNS allows its users to make full use of such features without having to incur the expenses of setting up, managing, and maintaining their own data network.

Electronic Data Interchange

Electronic data interchange (EDI) is the electronic transfer of commercial documentation between businesses and offices. The documents may include such items as purchase orders, delivery notes, invoices and receipts. The employment of EDI by a business cuts down enormously on paper work and so it results in a considerable increase in the speed of preparing and sending orders, etc., and cuts out the delays associated with the sending of paperwork through the post. The BT system is known as EDI*Net. Some examples of EDI that are in use in the UK are as follows.

(*a*) Brokernet is a system that links the offices and the computers of insurance brokers and insurance companies. It allows brokers and insurance companies to transmit documents electronically between their respective offices with a dramatic increase in speed.

(*b*) Tradanet is a network which links together many manufacturers, wholesalers, distributors and retailers in the retail field. The network is used to transmit orders, invoices, receipts, etc., from retailer to wholesaler, from wholesaler to manufacturer, and so on.

(*c*) Drugnet links a large number of doctor's practices to drug companies to enable them quickly to obtain information about various drugs.

(*d*) Factornet is used by firms, generally fairly small ones, that use factors to obtain payment of their bills, etc.

BT provide an EDI to FAX facility which allows users to transmit EDI documents to non-EDI customers who, however, do have a FAX machine.

Other VANs are many and include such diverse services as electronic publishing, computer-controlled movement of goods, direct marketing, security systems, telephone conferencing, stock ordering, industry, product and trade information, news, sport, information and weather forecasts, directories and word processing.

Automated Telling Machine Network

An automated telling machine (ATM) network connects a large number of cash dispensing machines to a data network and hence to a main computer. An example in the UK is the LINK network which incorporates banks, building societies and some other financial institutions in one operation. The principle of the LINK system is shown by Fig. 7.4. A request for cash at, say, a cash dispenser of building society A is sent to that building society's main computer. If the request for cash has been made by a customer of building society A that society's own computer will check the customer's account and, if the account holds sufficient money, the computer will send back

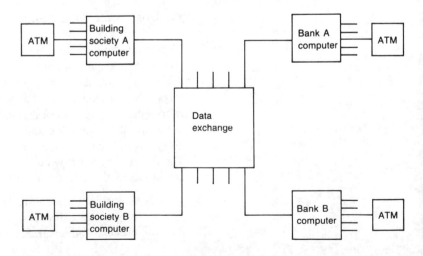

Fig. 7.4 ATM network.

a signal to the cash dispenser authorizing it to pay out the cash. If the cash request is made by a customer of another building society or a bank the request will be forwarded by the main computer to the main computer of the other organization for authorization. The individual bank and building society computers are connected to the data exchange by a 19.2 kbit/s link routed via either Kilostream or Megastream circuits. The computers of each organization are connected to their many cash dispensers using one of the data networks described in Chapter 6.

Electronic Point-of-sale Systems

An electronic point-of-sale (EPoS) system uses a front-entry data terminal, such as an electronic till, designed for use in a particular kind of shop or store. The data obtained provides the shop or store management with information on sales, remaining stock, turnover, re-ordering, etc. A terminal can be *stand-alone* with an external memory which acts as a program storage medium and a data carrier, or it can be a part of a master—slave system. With the latter system one terminal not only performs its own functions but it also supplies all the other *slave* terminals with their programs and data. In the larger shops, such as department stores and supermarkets, the EPoS terminals are connected to an internal network and will have either (*a*) a host terminal linked to the computer, or (*b*) a direct link to the computer via a communications controller (which may be a personal computer) and a modem.

An EPoS system incorporates price look-up tables which store information on the price and description of each product sold by the store; this allows the centralized pricing of goods, the description of stock items and detailed stock control. All that the till operator has to do is to identify the item being sold, enter its stock number and/or its price code, and then its full description and its price will be displayed on a VDU and, if required, this can be printed out. Weighing scales may be linked to the terminal and these can be used in conjunction with preset price per pound (or kilogram) keys. The input data to an EPoS terminal can be entered in different ways including hand or flat scanners to read bar codes, and light pens, manual keyboards, etc. Some terminals may also control other operations, e.g. in petrol stations and garages a terminal may be linked to a pump control and a tank monitoring gauge system. The advantages of an EPoS terminal are: (*a*) the speed of processing, (*b*) the accuracy of pricing and (*c*) the amount of information given on the print-out.

Electronic Funds Transfer At Point Of Sale

Increasing numbers of EPoS terminals incorporate *electronic funds transfer at point-of-sale* (EFTPoS), such as the SWITCH system of the Midland Bank. EFTPoS is the electronic retail and banking

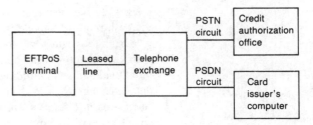

Fig. 7.5 EFTPos network.

network that allows the cost of the goods bought in a store to be automatically debited from the customer's bank account. The customer presents a plastic card which, when inserted into the EFTPoS terminal, authorizes the transfer of the money from the customer's bank account to the bank account of the store. The EFTPoS terminal may be operated either off line or on line; off-line operation means that data on all the transactions made is stored by the terminal for transmission later on. This transmission may be carried out electronically at night-time when communication costs are lower than in the daytime, or in some cases by hand or by post. On-line operation means that the EFTPoS terminal has instant access to the host computer and the money can be debited from the purchaser's account much more quickly; this access is often given by BT's Cardway system, the basic idea of which is shown by Fig. 7.5

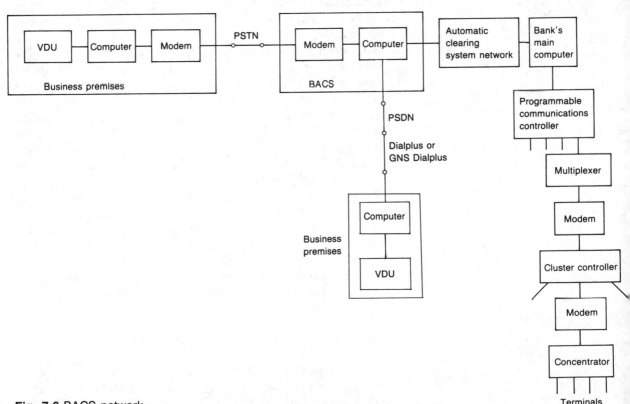

Fig. 7.6 BACS network.

Electronic Funds Transfer

Within the UK the transfer of money between the branches of a bank, and between different banks, is handled by a system known as the *Banker's Automated Clearing Service* (BACS). The BACS system is also used by many organizations for the payment of the salaries, pensions, etc., of their present and former staff into their individual bank or building society accounts. Some firms also use the system for the payment of accounts, etc. The basic block diagram of the system is shown by Fig. 7.6. The connection between the firm's computer and the BACS computer is usually dial-up via the PSTN or the PSDN unless the volume of data traffic warrants the cost of a leased circuit, analogue or digital, or perhaps a packet-switched connection.

8 Protocols

A protocol is a set of rules by which two *stations* (computer or terminal) may transfer data in a number of blocks or frames between one another. A protocol includes rules for the synchronization of the clock in the receiver, for determining which station is in control of the link, for the detection of errors, and for the maintenance of data flow. A protocol should ensure that garbled data does not look like a synchronization pattern. There is often more than one level of protocol in a network. The lowest level is concerned with hardware where a set of rules are necessary to specify how data can be transferred from a computer or terminal to the line and vice versa. This means that the ITU-T V24 and the EIA 232 computer–modem interfaces are forms of protocol, and so are the ITU-T X21 recommendations for the interface to a digital network. These low-level interfaces do not provide any form of error detection. A higher-level protocol ensures that the data is received at the far end of a circuit without error, and hence it does include error correction. In addition, a higher-level protocol also gives *data flow control*; this means that it ensures that the flow of data through the network is smooth, that no overloading occurs anywhere within the network and that the sending station does not send frames faster than the other station is able to receive them.

Most equipment manufacturers have their own protocols and these are usually unable to work with any other protocol. Hence, the two stations at each end of a point-to-point link *must* both employ the same protocol. Figure 8.1 shows a point-to-point link that connects two computers together. The two computers must both transmit their data

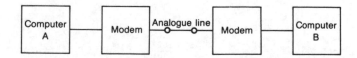

Fig. 8.1 Basic computer-to-computer link.

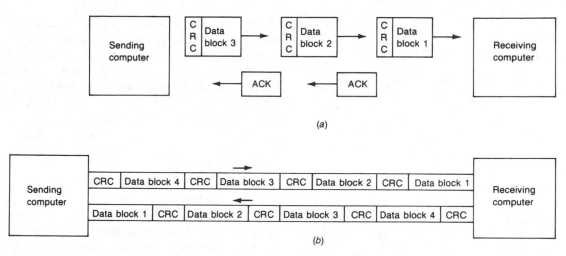

Fig. 8.2 (a) Half-duplex and (b) full-duplex protocols.

either synchronously or non-synchronously, at the same bit rate, using either a half-duplex, or a full-duplex, protocol. Figures 8.2(a) and 8.2(b) show respectively the essential principles of a half-duplex protocol, and a full-duplex protocol. In the half-duplex system each block of transmitted data must be acknowledged by the receiver before the next block of data is sent but in the full-duplex case this is not so. Both systems could be operating at the same bit rate and sending data in blocks of equal length but the full-duplex protocol will give the greater throughput. If a half-duplex protocol is employed some kind of line control is necessary to make sure that both computers do not attempt to transmit data at the same time. Usually, this is arranged by placing one computer in control of the link; the controlling computer will then poll the other computer to check whether it has any data to transmit and/or it is ready to receive data. This method of line control involves an *overhead* which reduces the overall data transmission efficiency but it is the approach used for such protocols as HDLC and SDLC. The overhead can be reduced if, when there is no data being transmitted, neither computer is in control of the link. Then when either computer has data to send it can contend for the link and when it has assumed control it can transmit its data. At the end of the transmission the computer will relinquish its control of the link and the link once again becomes available for either computer to contend for link control. The computer that has taken control is said to be the *master* station and the other computer is then the *slave station*. This system is used by the BiSynch protocol. If two different makes of computer are at one end, and several terminals are at the other end of a link, communication between them can be arranged using multiplexers. A protocol is then necessary to control the system. The more complex is the protocol the higher will be its cost but the greater will be the possible savings in both line and equipment costs.

Protocols are classified according to their method of framing and they fall into three main groupings.

(a) *Character-orientated protocols* employ special characters to separate the different segments of the transmitted information frame. The most commonly employed example of this type of protocol is BiSynch. Character-orientated protocols are inflexible since all messages are transmitted as a series of bytes. Sometimes data words may be of different lengths and then some bytes may contain only one, or two, actual data bits with the byte completed by the addition of padding bits. Binary data is difficult to handle since some data may appear to be a control code.

(b) *Byte-count protocols* employ a header that includes a count field which indicates the number of characters to come and the number that have already been received without error. Within the count field any character may occur and will not be taken as a control character. The main example of this type of protocol is DEC's DDCMP. The format of a DDCMP block of data is shown by Fig. 8.3.

(c) In *bit-orientated protocols* each frame is made up of a field in between eight-bit start and stop flags. The bits of each field, except the information field, are encoded with address, control, count, and error-checking bits. The data does not have to be arranged as a sequence of bytes but can be sent with any bit pattern.

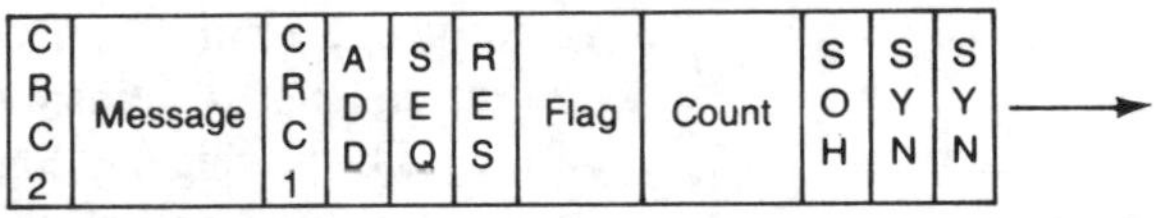

Fig. 8.3 DDCMP format (SEQ = sequence; RES = response).

Two modern protocols are *synchronous digital hierarchy* (SDH) and *asynchronous transfer mode* (ATM), and these provide flexible standards for both voice and data communications. The protocols are divided into three levels of hierarchy, (a) intra-office, up to 2 km, (b) inter-office, 2 to 15 km, and (c) long-haul, over 15 km.

BiSynch Protocol

The binary synchronous (BiSynch) protocol allows synchronous data to be transmitted in blocks each of which is preceded by a synchronizing sequence of bits which is usually the ASCII character SYN. BiSynch can only be used to provide synchronous half-duplex operation over either point-to-point or multi-drop circuits that are either two- or four-wire presented. The SYN characters are used by the receiving station to achieve character synchronization. For the receiver to achieve character synchronization it must detect two consecutive SYN characters in the incoming bit stream. Once the receiver has been character synchronized the rest of the received data is divided into eight-bit characters. As each block of data is received the receiver sends back to the transmitter an acknowledgement that

| B C C | E T B | Message | S T X | E O H | Header | S O H | S Y N | S Y N | Direction of travel → |

Fig. 8.4 BiSynch protocol.

the data has, or has not, been received without error. If zero errors were detected the positive acknowledgement ACK is sent; if an error was detected the negative acknowledgement NAK is sent. Any block of data that produces the NAK response from the receiving station is retransmitted by the sending station. Essentially therefore, the protocol operates on a half-duplex basis and the modems introduce a turnaround delay that may often be unacceptable. Therefore, a link is often operated using a four-wire presented line in order to minimize the turnaround delay and so obtain a greater throughput.

The format of the BiSynch protocol is shown by Fig. 8.4. Two SYN characters are followed by the start-of-header character (SOH) and then the header. This header may, or may not, be followed by the end-of-header character (EOH) before the start-of-text character (STX) signals the start of the message block itself. The end of the message is indicated by the end-of-transmission-block character (ETB), or if the data block is the last block in the message by the end-of-text (ETX) character. The header is not always present but when it is it gives such information as station control and priority. Each block of data, except the last block, has its end indicated by the end-of-transmission-block (ETB) character, but the final block is terminated by the end-of-text (ETX) character. Each transmitted character is checked for error and as each block of data is completed the block-check character (BCC) is transmitted. Finally, the end of data transmission is signalled by the end-of-transmission (EOT) character. If a block of data is received without error the receiving station transmits ACK0 and ACK1 alternately so that there is a running check to make sure that each acknowledgement corresponds to the immediately preceding block of data. The ACK0 signal is actually DLE0 and the ACK1 signal is really DLE1, where DLE is the data link escape character. If an error has been detected the receiver will return the NAK character to the sending station and the entire block of data will be retransmitted. Should further NAK characters be returned the transmitter will take the link as being faulty and it will stop sending data.

With a point-to-point link each station contends for control of the link whenever it has data to send; it does this by transmitting the control character ENQ. In a multi-drop network each station is given a unique address and once it has been addressed it indicates its readiness to receive, or transmit, data by sending the acknowledgement character ACK0. The slave stations in a multi-drop network idle in a *synch-search mode* waiting to identify the reception of two consecutive SYN characters. When the two SYN characters are received a character available flag is raised and the clock in the receiver is synchronized with the clock in the sending station. After a small delay the remainder of the protocol format follows.

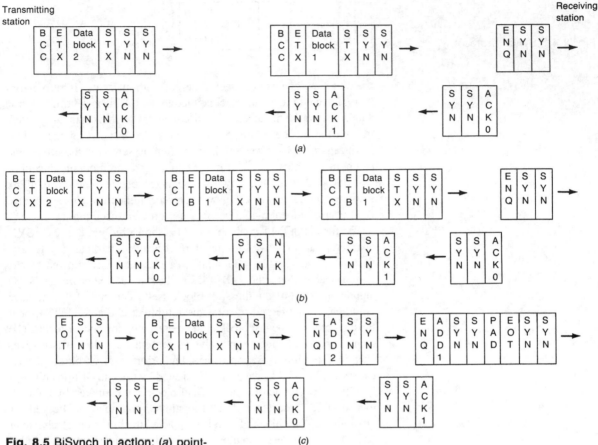

Fig. 8.5 BiSynch in action: (a) point-to-point link without error; (b) point-to-point link with error and (c) polled network.

Some examples of the BiSynch protocol are shown by Figs. 8.5(a), 8.5(b) and 8.5(c). In Fig. 8.5(a), which refers to a point-to-point link, the control station sends ENQ and when this is read by the receiving station it returns ACK0. When this acknowledgment arrives at the transmitting station it sends its data which, in this example, is two blocks in length. Each block is received without error and is acknowledged by the receiver transmitting first ACK1 and then ACK0.

In Fig. 8.5.(b) the first block of data received is not error free and so the receiving station returns NAK to the sender. This first block of data is then retransmitted and this time it is received without error and so the receiving station returns ACK1. Now the control station is able to transmit its second block of data and its correct reception is signalled by the receiver by returning ACK0.

Figure 8.5(c) shows a possible sequence of events in a polled network. After two successive SYN characters have been transmitted, first the EOT character to reset the network, and then the second (an idle PAD) character, which consists of seven 1s, are transmitted. These characters are followed by the ENQ character which asks the

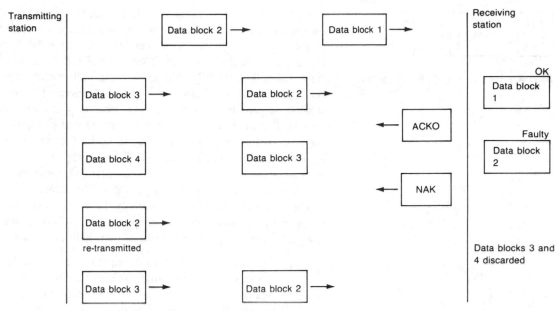

Fig. 8.6 ARQ system.

polled network whether it is ready to receive data. The first terminal polled, whose address is ADD1, is *not* ready to receive data and so it sends EOT along with the PAD and SYN characters. The computer then polls the next terminal whose address is ADD2; this terminal is ready to receive and so it returns ACK0 to the computer. The computer is now able to transmit the message which, in this case, is only one block in length. The correct reception of this message is acknowledged by the receiving station transmitting ACK1 and when the computer receives this it terminates the transmission by sending EOT. If the transmitting station does not receive an acknowledgement within a short time interval the reply request character (ENQ) will be transmitted which requests the receiving station to retransmit its last acknowledgement.

In the basic automatic repeat request (ARQ) system just described the sending station transmits a single block of data and then waits for its correct reception to be acknowledged. A more complex system allows blocks of data to be transmitted without waiting for an acknowledgement. If any data block is received with an error the receiving station will return the NAK character to the sending station and, at the same time, will discard all further blocks received until the faulty block has been received correctly. When the NAK character arrives at the sending station the station will retransmit both the faulty block of data and all the other blocks that had been sent after it. The basic principle of this ARQ system is shown by Fig. 8.6.

If the transmitted information includes any of the special control characters they must be made *data transparent*. This means that the characters must not be recognized as control characters by the receiver. A block of data can be made to appear transparent to the

B C C	E T X	D L E	Data block	S T X	D L E	S Y N	S Y N

Fig. 8.7 Use of the DLE character.

control circuitry by opening and closing the block with the data link escape character DLE. DLE is inserted just before the STX character at the beginning of the block and just before the ETB, or ETX, character at the the end of the block (see Fig. 8.7). In the message field DLE will then be the only control character that will be recognized as such; if any other control character is to be executed it must be preceded with DLE. The receiver will then remove the first DLE character it receives and treat the rest of the block, until the next DLE arrives, as pure data. If the DLE character itself is to appear in the message then another DLE character must be inserted in front of the text DLE, the receiver will then remove the first DLE and treat the second DLE as a pure data character.

The procedure outlined is essentially the method used by IBM and other manufacturers may employ slightly different versions of BiSynch. The BiSynch protocol has two main disadvantages.

(a) The need for each block of data to be acknowledged before the next block is transmitted means that the protocol is inherently half-duplex in its operation and this considerably reduces the throughput of the system.
(b) The DLE character must be used to give complete message transparency.

These disadvantages can be overcome by the use of a bit-orientated protocol such as *high-level data-link control* (HDLC), *synchronous data-link control* (SDLC) and ITU-T X25. HDLC is the International Standard Organization's protocol and SDLC is essentially one version of it and in this book these protocols will be taken as being the same except where otherwise mentioned. X25 is really another version of HDLC applied to gaining access to a packet-switched network.

High-level Data-Link Control

High-level data-link control (HDLC) is a set of protocols for use with wide area networks (WANs) that largely overcomes the disadvantages of character-orientated protocols, such as BiSynch, i.e. only a half-duplex protocol and the need to use the DLE character to obtain message transparency. The two main protocols in the HDLC set are LAPB for point-to-point links and RNM for multi-drop links.

HDLC is a full-duplex protocol (although it can be used in a half-duplex mode) which means that a link uses two independent channels and synchronous transmission. While pure binary messages, i.e. not separate characters, are being transmitted over one channel, acknowledgements can be transmitted over the other channel in the reverse direction. The sending station transmits blocks of data continuously and only stops if it receives notification of an error in a particular block. By the time the NAK signal has been received several other blocks after the faulty one will have been transmitted. The transmitted blocks of data must therefore be numbered so that they can be

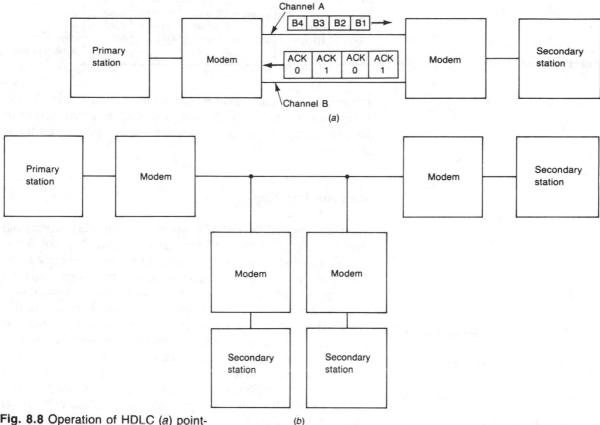

Fig. 8.8 Operation of HDLC (*a*) point-to-point link and (*b*) polled network.

identified individually, each block must also be stored at the transmitter for the time required for any error notification to be received.

In a multi-drop network it is possible for the computer to transmit simultaneously to one terminal and receive from another. One station in a link is permanently made the master. It is responsible for the control of the link and it is known as the *primary station*. The other station(s) in the network are called the *secondary stations* and they can only respond to commands issued by the primary station. Two examples of this are shown by Figs. 8.8(*a*) and (*b*). In Fig. 8.8(*a*) two stations are connected together by a point-to-point link. The primary station will be a computer while the secondary station may be either a computer or a data terminal. In Fig. 8.8(*b*) the primary station is a computer while all three secondary stations are terminals; four-wire presented circuits are employed in both networks.

Frame Structure

Figure 8.9 shows the format of a HDLC frame; the start flag, the address field and the control field are known as the *header*. A trans-

Eight-bit stop flag	16-bit frame check sequence	Message	Eight-bit control field	Eight-bit address field	Eight-bit start flag

Direction of transmission →

Fig. 8.9 HDLC frame.

mitted frame will contain either supervisory or message data. Supervisory frames are used for the confirmation of the correct receipt of information frames, ready and busy conditions, and to report frame sequence errors.

Start and Stop Flags

The beginning and the end of a message are signalled by start and stop flags that consist of the bit pattern 01111110. The start flag is also used to derive the synchronization of the receiver clock to the clock in the transmitter. All the active secondary stations search for this flag so that they can synchronize with it. Note that when two, or more, frames follow immediately after one another only one flag is needed since the stop flag of one frame will also act as the start flag of the next frame. To preserve information field transparency this bit sequence is not allowed to occur in the information field; if it should do so the transmitter will insert a 0 after the fifth 1 (this is known as 'bit stuffing'). If the receiver detects five consecutive 1s followed by a 0 it will remove this 0 to restore the data to its correct form. This is shown by Fig. 8.10.

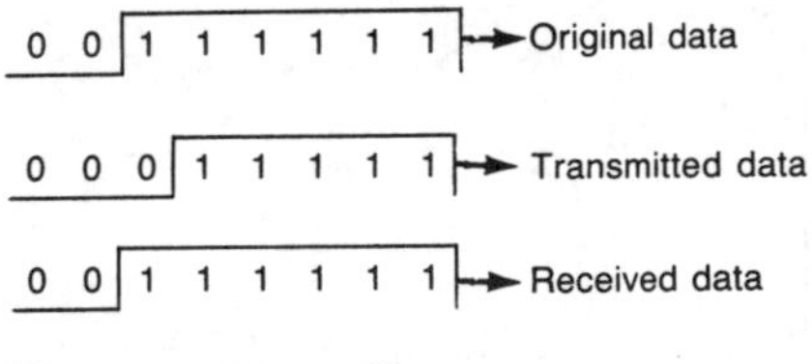

0 0 1 1 1 1 1 1 → Original data

0 0 0 1 1 1 1 1 → Transmitted data

0 0 1 1 1 1 1 1 → Received data

Fig. 8.10 Bit stuffing.

Address Field

The eight-bit (sometimes 16-bit) address field identifies the destination secondary station; it is, of course, not really needed in a point-to-point circuit but it is always included. When the primary station sends to a network the address field identifies the wanted secondary station. When data transmission is in the other direction the address field identifies the secondary station to the primary station. The primary station itself does not have an address. The address 11111111 is used to indicate an all-station broadcast.

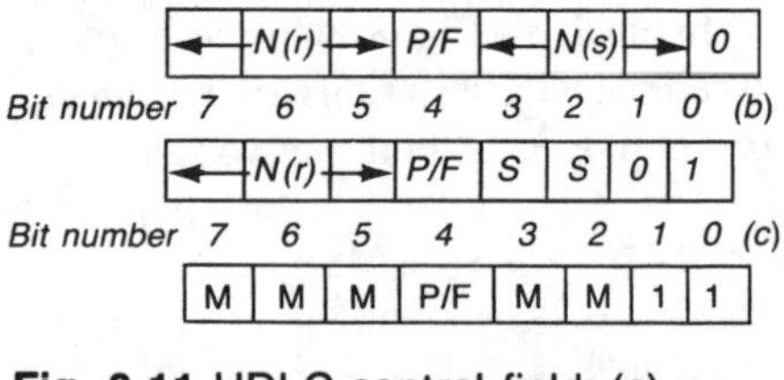

Bit number 7 6 5 4 3 2 1 0 (a)

←	N(r)	→	P/F	←	N(s)	→	0

Bit number 7 6 5 4 3 2 1 0 (b)

←	N(r)	→	P/F	S	S	0	1

Bit number 7 6 5 4 3 2 1 0 (c)

M	M	M	P/F	M	M	1	1

Fig. 8.11 HDLC control field: (a) information, (b) supervisory and (c) un-numbered.

Control Field

The eight-bit (sometimes 16-bit) control field, which defines the function of the frame, is in one of three formats: supervisory, information and un-numbered. These three formats are shown in Fig. 8.11. Bit 0 identifies the frame as being either an information frame or a command/response frame. There are two kinds of command/response frame, namely the supervisory frame and the un-numbered frame. The supervisory frame is used to initiate and control the transfer of information and the un-numbered frame is used to set the operating modes and also to initialize all the stations. In all three kinds of frame the P/F bit is used to identify whether a frame is (a) being sent from a primary station to a secondary station, when P/F = 1, or (b) is being sent from a secondary station to the primary station, when P/F = 0, but the P/F bit is always set to 0 for the final response frame.

Information Frame

The information frame is used to transfer information and it has its bit 0 set to 0. $N(s)$ and $N(r)$ are the send and receive sequence counts respectively (0 to 7), and they are maintained by each station for the information frames sent and received by that station. In a polled network, each of the secondary stations maintains its own $N(s)/N(r)$ count but the primary station maintains a separate count for each of the secondary stations. The receive sequence count advises the other station of the expected sequence number of the next frame to be received, and hence it acknowledges that all the previous frames have been received without error. P/F is the *poll/final* bit that is used by the primary station — when set to 1 — to request a response from a secondary station, i.e. it acts as a poll. The response may be a single frame or it may consist of several frames. A secondary station generally uses the P/F bit set to 1 to indicate the last frame in the sequence of frames that it is sending. The P/F bits are always exchanged between the primary station and the secondary station in pairs. The information field is usually some multiple of eight bits in length.

Command/Response Frame

The command/response frames are used in the control of data transfer over a link. A command can only originate from a primary station and a response can only originate from a secondary station. The formats of the supervisory and un-numbered frames are shown by Figs. 8.11(b) and (c).

Supervisory

A supervisory frame is used for flow and error control, it confirms the reception of information frames, conveys ready or busy signals, and reports errors. If bit 0 is set to 1 it identifies the frame as a command/request frame, and if bit 1 is set to 0 it identifies the frame as a supervisory frame. The information field is then not present. Bits

Table 8.1

0	0	Receiver ready (RR)	All frames up to $N(r)-1$ received correctly.
1	0	Receiver not ready (RNR)	All frames up to $N(r)-1$ received correctly. Send no more frames until RR is returned.
0	1	Reject (REJ)	Re-transmit starting with frame $N(r)$.
1	1	Selective reject (SREJ)	Re-transmit specific frame number $N(r)$.

2 and 3 may have any of the values given in Table 8.1. The SREJ option is often not implemented. RR and RNR are very similar to ACK and NAK in BiSynch.

The P/F bit acts in the same way as in the information frame, i.e. as a poll when set to 1 by the primary station, and as an end-of-message indicator when set to 1 by a secondary station. Bits 5, 6 and 7 contain the $N(r)$ receive sequence count which allows the receiving station to acknowledge the correct receipt of all the previous frames.

Un-numbered

The un-numbered frames provide five bits; known as modifiers (M), that can be used to provide extra commands and responses, and details are given by Table 8.2.

Information Field

In HDLC the information field can be of any length but in SDLC its length must be a multiple of eight bits. In each byte the least

Table 8.2

Bit number	7	6	5	4	3	2	1	0	Function
Commands	0	0	1	P/F	0	0	1	1	Un-numbered poll
	0	0	0	P/F	0	0	1	1	Un-numbered information
	1	0	0	P/F	0	0	1	1	Set normal response mode
	0	1	0	P/F	0	0	1	1	Disconnect
	0	0	0	P/F	0	1	1	1	Set initialization mode
	1	0	1	P/F	1	1	1	1	Exchange station identification
	1	1	1	P/F	0	0	1	1	Test
	1	1	0	P/F	0	1	1	1	Configure
Responses	0	0	0	P/F	0	0	1	1	Un-numbered information
	0	1	1	P/F	0	0	1	1	Un-numbered ACK
	0	0	0	P/F	0	1	1	1	Request initialization mode
	0	0	0	P/F	1	1	1	1	Disconnect mode
	1	0	0	P/F	0	1	1	1	Frame reject
	1	0	1	P/F	1	1	1	1	Exchange station
	1	1	1	P/F	0	0	1	1	Test
	0	1	0	P/F	0	0	1	1	Request disconnect
	1	1	0	P/F	0	1	1	1	Configure

significant bit is transmitted first. The contents of the field are treated as binary data even though it may consist of ASCII characters.

Frame Check Sequence

The 16-bit frame check sequence checks the received data for errors using a 16-bit *cyclic redundancy check* (CRC) based upon the ITU-T recommendation V41. It uses the generating polynomial $X^{16} + X^{12} + X^5 + 1$. The block check character computed from the address, control and information fields make up the frame check sequence. If a received frame is error free the receiver count $N(r)$ is advanced by 1, if it is in error the count $N(r)$ is not advanced.

Full-duplex Point-to-point Data Transfer

There are two modes of operation allowed with HDLC that are known, respectively, as the *normal response mode* and the *non-synchronous response mode*. Since the former is very much the more commonly employed only it will be considered. In this mode of operation a secondary station can only transmit data after it has responded to a poll from the primary station. Figure 8.12 shows a typical sequence of transmitted signals all of which are received without error. It has been assumed that data is only sent from the primary station to the secondary station. The sending sequence reverts to 0 after block 7 because this is the maximum number of blocks that can be transmitted without an acknowledgement. If a transmitted information frame was faulty the acknowledgement returned to the sending station would indicate the faulty frame. If, for example, frame 2 was received

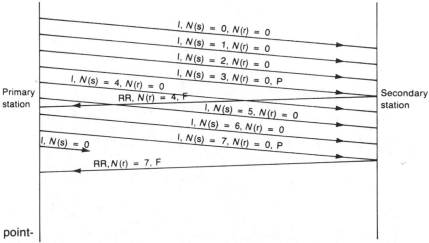

Fig. 8.12 Signal sequence in a point-to-point data transfer.

incorrectly then the response would be RR, $N(r) = 2$, F, to indicate that the last correctly received frame was number $N(r) - 1 =$ frame 1.

ITU-T V42

The ITU-T recommendation V42 is a full-duplex protocol that has two parts. Part 1 is really MNP IV (page 157), and it acknowledges the existence of a large number of systems that use that protocol; part 2 is a development of the ITU-T X25 Link access procedure-balanced (LAP-B) protocol known as Link access procedure for modems (LAP-M). V42 bis deals with data compression in addition to V42 error correction for non-synchronous data transmission. V42 bis is able to recognize that a file has been previously compressed, and if it has will not attempt further compression.

9 Error Control

Error detection and, less often, error correction, is employed on data circuits to overcome the corruption and/or the loss of information of the data signals arriving at the receiving end of a link. With the simplest form of error detection a parity bit is added to the end of each eight-bit ASCII character. The parity bit is used to make the number of 1s in each character *either* an odd number — known as odd parity — *or* an even number — known as even parity. The parity bit makes it possible for the receiver to detect a single error in a character but not two errors. Except for the simplest systems the transmitting terminal must be informed if an error has occurred in the received data so that the data can be retransmitted. To make this possible the data is broken up into a number of blocks and each block has a number of block check characters added to it. The receiver recomputes these block check characters from the received data and uses the result to determine whether or not the data is free from error. If no errors have been detected the correct receipt of the data is acknowledged, by returning the ACK character to the sending terminal, and the next block of data is transmitted. If an error has been detected this fact is signalled back to the sending terminal by the character NAK and then the sending terminal will send that block of data again. This is often known as an *automatic repeat request* (ARQ) system and it is used in conjunction with bit-orientated protocols such as BiSynch.

A more powerful method of error detection, known as a *cyclic redundancy check* (CRC), is used with the HDLC protocol. A number of different CRC systems are available but most HDLC links follow the ITU-T V41 recommendations.

If extra redundant bits are added to the data it is possible also to correct a detected error at the receiver. Because of the need for the extra redundant bits such a system, known as *forward error control*, is only employed when a return channel, over which acknowledgement of the correct, or incorrect, receipt of data can be signalled, is not available.

Parity Error Control

There are two forms of parity error control system, known as *character parity* and *block parity*.

Character Parity

When character parity is used an extra bit is added to each character in the data message. The ITU-T recommend (V4) that the parity bit is placed after the seventh and most significant bit and so it becomes the eighth bit. For example, the character 1010000 would have a 0 parity bit added in an even parity system, and a 1 parity bit added if odd parity is used. Thus the transmitted character would be either 01010000 or 11010000. It is customary to employ even parity on non-synchronous circuits and odd parity on synchronous circuits. The receiver checks the received characters and if the total number of 1s is even for even parity, or odd for odd parity, it assumes that the data is error free.

The parity bit can be generated using either hardware or software and one simple hardware circuit is shown by Fig. 9.1. It can be seen to employ six exclusive-OR gates and, perhaps, one inverter; the gates would require two quad EXOR IC packages such as the SN7486 or the SN74386. Suppose that the seven-bit word applied to the circuit is 1010000; then, since for an exclusive OR gate $1 + 1 = 0$, $0 + 0 = 0$, $1 + 0 = 1$ and $0 + 1 = 1$, Table 9.1 shows the output logical state of each gate. The output of gate F is the required even parity bit, it can be seen that this is 0 as is required for even parity. If odd parity is wanted then the inverted output is used.

The same circuit, with one extra gate, can be used at the receiver to generate a parity bit from the eight-bit received character. If there is no single error in the received character Table 9.1 will still apply

Table 9.1

Gate	Inputs	Output
A	00	0
B	00	0
C	00	0
D	10	1
E	01	1
F	11	0

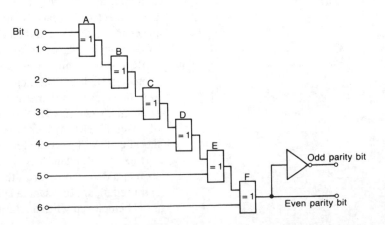

Fig. 9.1 Generation of the parity bit.

Table 9.2

Gate	Inputs	Output
A	00	0
B	00	0
C	00	0
D	10	1
E	11	0
F	10	1
G	01	1

Table 9.3

Gate	Inputs	Output
A	00	0
B	10	1
C	01	1
D	11	0
E	10	1
F	11	0
G	00	0

Bit number	7 6 5 4 3 2 1 0
	0 1 0 0 0 0 0 1
	1 0 1 0 0 0 0 0
	0 1 0 0 0 1 1 1
	1 1 0 0 1 1 1 1
	1 1 0 0 1 1 1 1
Characters	1 0 1 0 0 0 0 0
	0 1 0 0 0 0 1 0
	0 1 0 0 0 1 1 1
	0 1 0 0 0 1 1 1
	0 1 0 0 1 0 1 1
BCC	0 0 0 0 1 1 1 1

Fig. 9.2 Derivation of the BCC.

with the addition of a gate G which will have inputs of 0 and 0 and hence an output of 0. If any *one* of the received bits, say bit 5, is incorrect the output of the receiver's circuit will be at logical 1 indicating the presence of a error, see Table 9.2. If, however, two errors were present no indication of error would be given. This is shown by Table 9.3 in which bits 3 and 6 of the received signal are incorrect, i.e. the received signal is 01110100.

Block Parity

The efficiency of error detection can be considerably increased by the use of block parity. The data message is divided into a number of blocks and each block has a *block check character* (BCC) added at the end of the block. Figure 9.2 shows a block of 10 characters that has even parity applied to each character. The parity bits at the end of each character provide a *longitudinal redundancy check* (LRC). The BCC is the even parity of each of the individual character bit columns, known as the *vertical redundancy check* (VRC), and in this case it is equal to 0001111; this, in turn, has even parity applied to it and so the complete BCC is 00001111. The BCC is transmitted at the end of the block of data.

At the receiver the parity of each row and of each column is checked and the location of any single error is indicated by the intersection of the column and the row containing the error. If there are two errors in one character the character parity bit will be correct but the BCC will be incorrect; this means that the presence of an error will have been detected but not its location within the block. Similarly, if there are two errors in the same numbered bit of two separate characters the BCC will be correct but the parities of the two characters will be incorrect. Not all patterns of bit errors will be detected.

The generation of the BCC is not based upon all of the characters in a block of data. This is shown by Fig. 9.3 for two different cases (see Fig. 8.4 also). The generation of the BCC commences after the first appearance of either the SOH (start of header) character or the STX (start of text) character but this character is not itself included in the computation. If, however, the data block starts with SOH followed by STX the STX character will then be included in the

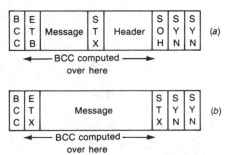

Fig. 9.3 BCC computation.

calculation of the BCC. At the end of the block of data, detection by the receiver of either the ETB (end of block) or the ETX (end of text) characters indicates to the receiver that the next character to arrive is the BCC. By the time the BCC has been received the receiver will have generated its own BCC from the received data and so the receiver will be able to compare the two BCCs with one another. If they are not the same then an error has occurred somewhere in the block of data and the receiver will return NAK to the transmitting terminal. If the two BCCs are identical then no error has been detected and the receiving terminal will return ACK to the transmitting terminal.

The BCC can be generated using either hardware or software although the former method is the more commonly employed. Hardware generation of the BCC essentially consists of exclusive-ORing all the preceding characters received in a block of data.

```
0 0 0 0 0 1 0
1 0 1 0 0 0 1
1 0 1 0 0 0 0
1 0 1 0 0 1 0
0 1 0 0 0 0 0
1 0 0 0 1 1 0
1 0 0 1 1 1 1
1 0 1 0 0 1 0
0 1 0 0 0 0 0
1 0 1 0 1 0 0
1 0 0 1 0 0 0
1 0 0 0 1 0 1
0 1 0 0 0 0 0
1 0 0 0 0 1 1
1 0 1 0 1 0 1
1 0 1 0 0 0 0
0 0 0 0 0 1 1
```

Fig. 9.4

Example 9.1

An ASCII coded block of data is shown by Fig. 9.4. (a) Use Table 1.1 to decipher the message. (b) If even parity is used determine the parity bit of each character. (c) Determine the BCC. (d) If, at the receiver, (i) the fourth bit of the 10th data character, (ii) both the fourth and the fifth bits of the 10th data character, and (iii) the fourth bits of both the 10th and the 12th data characters are incorrect, compute the BCC at the receiver.

Solution

(a) The message is QPR FOR THE CUP.   (*Ans.*)
(b) The parity bits are (first character on the right),
00011101111111011.   (*Ans.*)
(c) The BCC is 00110110.   (*Ans.*)
(d) The BCC is (i) 10111110, (ii) 00101110, (iii) 00110110.   (*Ans.*)

Cyclic Redundancy Checks

Data systems that use an HDLC protocol employ some form of *cyclic redundancy check* for the detection of errors. With an HDLC protocol each block of data is sent as a long binary number and it is not sent as a number of separate characters. Parity checking is then no longer possible.

At the transmitting terminal the binary number representing the data to be transmitted is first divided by a predetermined, and constant, number using *modulo-2* arithmetic. The division process produces both a quotient and a remainder; the quotient is not wanted and it is discarded but the remainder is used as the cyclic check code (CRC). The CRC is transmitted to the receiving terminal immediately after the block of data. At the receiver the incoming data, including the CRC, is divided by the same number as that employed at the transmitting terminal and if the received block is error free the remainder will be zero. The use of cyclic redundancy checks on synchronous data links is very efficient for the detection of errors. The binary number that is used as the divisor is often known as the *generating polynomial* and it should be one bit longer in length than the required length of the CRC. If the CRC is n bits long the data binary number should be multiplied by 2^n, i.e. n 0s added to the least significant end of the number. The ITU-T recommendation V41 requires a 16-bit CRC and it uses the generating polynomial $X^{16} + X^{12} + X^5 + 1$ or 10001000000100001.

Modulo-2 Division

With modulo-2 division of one binary number by another the rules are as follows.

(a) If the divisor has the same number of bits as the dividend the quotient is 1; if the divisor has fewer bits than the dividend the quotient is 0.

(b) There are no carries and $1 - 1 = 0, 0 - 0 = 0, 1 - 0 = 1$ and $0 - 1 = 1$. Since the division is binary any remainder will always be one bit shorter than the divisor.

As an example of the procedure consider the data number to be 1010110101 and suppose that the CRC is to be four bits in length. The divisor must then be five bits long and it must also have both its least, and most, significant bits equal to 1. Let the divisor be 11001 or $X^4 + X^3 + 1$. The message must be multiplied by 2^4, i.e. four 0s added to its least significant end. Thus

```
                    1100000110
11001 ) 10101101010000
            11001
            1100
            11001          bring down next bit (1)
            11001
                010100  bring down, in turn, bits 010100
                11001
                1101
                11010      bring down next bit (0)
                11001
                0110      remainder
```

The CRC can be generated, at both the transmitting terminal and the receiving terminal, by a suitable combination of shift registers and exclusive-OR gates. The number of stages in the shift registers must be equal to the required length of the CRC and the number of gates is one less than the number of terms in the generating

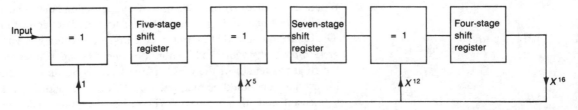

Fig. 9.5 Generation of the ITU-T CRC.

polynomial. Thus for the polynomial $X^4 + X^3 + 1$ four shift register stages and three exclusive-OR gates are necessary. The circuit that generates the ITU-T polynomial, i.e. $X^{16} + X^{12} + X^5 + 1$ is given in Fig. 9.5. Initially, all the shift register stages are cleared. The data signal is then entered, bit-by-bit, starting with the most significant bit and the first bits move across the stages in the usual shift register manner. Thereafter feedback modifies the operation of the circuit. When all the data bits have been passed into the circuit the remainder is held in the registers and this is then passed on to the receiver at the end of the message. At the receiver a similar circuit has the CRC clocked into it and if there is zero error in the received data then the contents of the shift registers will all be logical 0.

Forward Error Correction

A forward error control system can be employed to both detect and correct any errors in the received data. Since retransmission schemes using either BCC or CRC are more efficient, forward error control is rarely used unless a return channel is not available. Extra bits are required so that the location, as well as the existence, of an incorrect bit can be pinpointed. The most commonly employed error-detection code is the *Hamming* code. This code uses parity check bits that are located in set positions within each block of data; these bits allow multiple parity checks to be carried out at the receiving terminal. The positions of the Hamming bits are given by 2^n, where n is an integer. Thus, the Hamming bits are at positions 1, 2, 4, 8, 16, etc., in each block of data. All the remaining positions are occupied by the actual data bits. Consider the bit stream 11000101100100; when the Hamming bits have been inserted this becomes 110X0010110X010X0XX. It is necessary to determine whether each Hamming bit, marked as X, is a 1 or a 0. To do this the position of each 1 bit in the data bit stream is noted and the binary value of each position is added, using modulo-2 arithmetic. (With modulo-2 addition an even number of 1s = 0 and an odd number = 1 and there are no carries.) In this case there are 1 bits in the positions 6, 10, 11, 13, 18 and 19. Adding these gives the result shown in Table 9.4.

Inserting these Hamming bits into the data bit stream gives 1100001011010100011. At the receiving terminal the binary values of the position of each 1 in the received bit stream are added, again using modulo-2 arithmetic. If the data has been received without error the sum will be equal to zero as shown by Table 9.5.

If, however, an error in a single bit has occurred the position of

Table 9.4

19	1	0	0	1	1
18	1	0	0	1	0
13	0	1	1	0	1
11	0	1	0	1	1
10	0	1	0	1	0
6	0	0	1	1	0
	0	1	0	1	1

Table 9.5

19	1	0	0	1	1
18	1	0	0	1	0
13	0	1	1	0	1
11	0	1	0	1	1
10	0	1	0	1	0
8	0	1	0	0	0
6	0	0	1	1	0
2	0	0	0	1	0
1	0	0	0	0	1
	0	0	0	0	0

Table 9.6

19	1	0	0	1	1
18	1	0	0	1	0
13	0	1	1	0	1
10	0	1	0	1	0
8	0	1	0	0	0
6	0	0	1	1	0
2	0	0	0	1	0
1	0	0	0	0	1
	0	1	0	1	1

If, however, an error in a single bit has occurred the position of the incorrect bit will be indicated by the result of the modulo-2 addition. Suppose, for example, that the 11th bit in the received data bit stream is incorrect, i.e. it is a 0 instead of a 1. The modulo-2 addition of the binary values of the positions of the 1s in the bit stream is then given by Table 9.6.

The result, 01011, is, of course, the binary equivalent of decimal 11 and so the faulty bit has been located and can be corrected by inverting it to become 1.

ITU-T V42

The ITU-T recommendation V42 refers to the use of non-synchronous-to-synchronous conversion techniques for the checking of errors at a receiving modem. The recommendation is a compromise between two error-correcting protocols, known as MNP (Microcom networking protocol) classes 3 and 4, and the HDLC based ITU-T LAP-M (link-access procedure for modems). The LAP-M protocol is specified in the main body of the recommendation and MNP is given in an appendix. The MNP standard had been in use for a number of years before the LAP-M standard was introduced and it is often used for high-speed data transfer over unconditioned telephone lines. The classes of MNP are as follows.

(a) Class 1 is a synchronous character-orientated half-duplex standard but it is no longer used.

(b) Class 2 is a non-synchronous character-orientated full-duplex standard in which bytes are transmitted in fixed length packets that contain an error checking word. The receiving modem must acknowledge correct receipt of each packet.

(c) Class 3 is a synchronous bit-orientated full-duplex standard with which, since no start or stop bits are required, the overhead is reduced by about 25%.

(d) Class 4 is an error-correcting protocol that also provides some data compression. It incorporates *adaptive packet assembly* to allow the modem to package data into packets that are then transmitted and error checked as a complete entity. Also included is *data phase optimisation*, which eliminates repetitive control bits from the transmitted data stream. Together these two techniques increase the throughput of a modem to about 120%.

Data is transmitted in packets whose size is varied according to the error rate on the telephone line, if the error rate increases then smaller data packets are assembled.

Classes 2–4 are in the public domain but classes 5–10 are licensed by Microcom.

(e) Class 5 provides data compression for non-synchronous data. It can compress by a factor as great as 2, which effectively doubles the data transmission rate.

(*f*) MNP 6 uses a technique known as *Universal Link Negotiation*, which allows two modems to start sending data at a low bit rate and then, once the overall capability of the circuit plus modems is known, switch to a higher bit rate.

(*g*) MNP 7 uses *Huffman* encoding to provide a more efficient data compression system than MNP 5. Compression ratios as large as 3 are sometimes possible.

(*h*) MNP 9 (there is no MNP 8) is used to reduce the overheads introduced by certain modem operations. The acknowledgement of the correct reception of a data packet is combined with the acknowledgement of the following packet, and if an error is detected only the data in error must be retransmitted, instead of, as is often the case, all the data following the incorrect data.

(*i*) MNP 10 is a set of *Adverse Channel Enhancements* which makes it possible for a modem to work better over a poor telephone circuit. The protocol allows two modems to make repeated attempts to set up a link, to optimize the packet size, and to use the highest bit rate possible. MNP 10 is used with data communications over cellular radio links.

Unfortunately, perhaps, the MNP and LAP-M standards are not compatible and this is why V42 offers a compromise between them. V42 bis incorporates a data-compression technique that is compatible with the MNP standard. With the V42 standard the calling modem initiates a *detection phase* by sending a particular sequence of characters. If the receiving modem is also a V42 modem it returns another set sequence of characters to confirm its capability and whether it wishes to operate in the error-checking mode. For error-controlled operation the two modems move into the *protocol establishment phase* and once this has been established data transfer commences. If the receiving modem is not a V42 modem it returns the calling modem's character sequence back and on receipt of this the calling modem will try to use MNP. If this attempt also fails the calling modem will then go into its non-error correcting mode of operation.

V42 bis provides a compression ratio of 4:1. When V42 and V42 bis are used together they can increase the speed of a V32 bis modem from 14 400 bits/s to 57 600 bits/s, i.e. by a factor of 4.

10 Local Area Networks

The term *local area network* (LAN) is usually applied to a data communication system that operates within a single building, or between several sites that are only separated by distances of up a few kilometres. A LAN can connect together a number of computers, mainframe or mini, or most likely PCs, and give them access to computers and peripherals, such as printers and hard discs. Computers and terminals are known as *stations*. The connections between stations are not made, as with a wide area network or WAN, over analogue or digital telephone lines obtained from the telephone administration (BT or Mercury in the UK) but over locally installed cable. The cabling may consist of twisted pairs, multi-pair cable, coaxial cable, or optical-fibre cable. Because the bandwidth limitations imposed by the public telephone network no longer apply and error rates are low, a LAN is able to operate at much higher bit rates and transmission speeds of up to 16 Mbit/s over coaxial cable, and 100 Mbit/s over optical-fibre cable, are not uncommon. Such high speeds are intended for data transfer between two computers, or for allowing different computers to share expensive resources such as a high-capacity hard disc store; many of the devices employed on the LAN will be unable to work at such a high speed.

A LAN provides a high-speed communication network for the interconnection of the computers and the various types of terminal that are located within a limited geographical area, such as an office building or a factory site. A LAN allows an organization to employ distributed processing of data using PCs that are able to access one another and/or a host computer. In addition it provides:

(a) *resource sharing* in which a large number of computers, workstations, etc., are able to access a small number of expensive peripherals, such as databases, hard-disc drives, printers and plotters;

(b) *information sharing* in which all application programs and data stored anywhere in the network are available to every user; and

(c) *network access* in which all users have access to WANs.

A LAN can offer support for such services as:

(a) file transfer and access, i.e. the movement of long blocks of data, such as text, from one data terminal to another;

(b) graphics;

(c) word processing;

(d) electronic mail;

(e) access to a main database and/or distributed databases; and

(f) distributed processing.

In some cases digitalized voice and video signals may also be transmitted but it is more usual to employ a digital PABX for this purpose.

A LAN may be either *baseband* or *broadband*. Baseband means that the data is transmitted directly over the LAN as a digital signal, probably having been encoded first, and each station has exclusive use of the medium for a short period of time. Broadband means that the data from different stations is transmitted at the same time but in different frequency bands using a process that is known as frequency division multiplex (FDM). In general, baseband LANS are used for systems that convey data signals only, and broadband systems are used to transmit a mix of data, speech and video signals. The former is very much the more commonly employed.

A baseband LAN is designed to allow all of its stations freely to exchange data with any other station and to use any of the common resources. Protocols have been developed to control the access of each station to the LAN. The protocol ensures that only one station is able to transmit data into the network at any one time. The data is first assembled into packets and then it is transmitted. Each packet consists of a series of bits which form a block into which all, or a part (depending on its length), of a message can be placed before it is transmitted into the LAN. The format of each packet is defined and it may include source and destination addressing information, flags, and a CRC. A number of different baseband LANs have been developed but only two of them are used to much extent. These two techniques are known as token passing and Ethernet.

The operation of a LAN is controlled by software known as the *Network Operating System* (NOS). The functions of a NOS are:

(a) To allow the operating system of a PC connected to the LAN to communicate with the servers.

(b) To make it possible for one LAN to be connected to another LAN.

(c) To provide security arrangements that prevent unauthorized access to files and files being over-written.

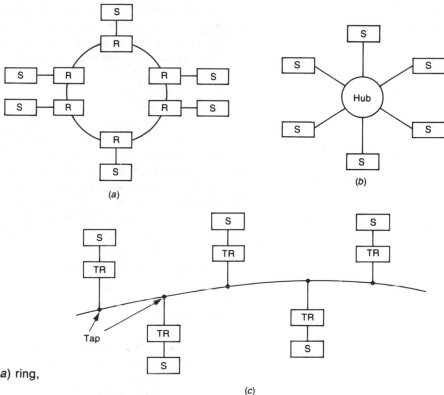

Fig. 10.1 LAN topologies: (a) ring, (b) star and (c) bus.

Topology

The topology of a LAN is the way in which the stations connected to the LAN are linked together. The three most commonly employed topologies are the ring, the star, and the bus networks and these three are shown in Figs. 10.1(a), 10.1(b) and 10.1(c), respectively.

Ring Network

A ring LAN consists of a number of repeaters connected in a continuous loop; each repeater is a node to which a station is connected. Data that is transmitted into the LAN by a station circulates around the ring in one direction only. A repeater is able to copy the circulating data from one segment of the LAN, regenerate it and then pass it on to either the next segment of the LAN or to its associated station if the data is addressed for that station. Data that is passed from the LAN to the station, or vice versa, will have its bit rate changed to the bit rate used by the LAN, or by the station. Typically, the LAN will operate at a high bit rate such as 10 Mbit/s while the speed of the station will depend upon its nature but may only be a few kilobits per second. A ring LAN is able to provide a high bit rate over distances of up to a few kilometres. The network requires active taps and will break down if any one of the nodes becomes faulty and/or there is a break in the cable. It is necessary to interrupt service whilst extra stations are connected into the ring or an existing station is removed. The ring network is used for token-passing LANs.

Star Network

A star network contains a central point, known as the hub, to which all the stations in the LAN are connected radially and through which all communications are routed. The star-connected LAN possesses the disadvantages of being relatively slow and, since the hub controls all switching between stations, being completely dependent upon the correct working of the hub for its operation. It is, however relatively cheap and easy to implement since it uses a relatively simple protocol. Generally, the star network is used in conjunction with a digital PABX but it is also used with fibre optic Ethernet systems.

Bus Network

A bus LAN consists essentially of an open loop to which are connected a number of interface nodes to which stations are connected. The interface consists of a cable tap and a transceiver. The transceiver conditions the signals passing to and from the cable so that the station can interface with the LAN. Two stations may, for example, be communicating with one another at a bit rate of 4.8 kbit/s. The data transmitted by one station is sent to the transceiver and here it is placed into temporary storage before it is transmitted on to the LAN in a short burst at a much higher bit rate, say 10 Mbit/s. The receiving transceiver stores the received bursts of data and then sends the received data to its associated station at 4.8 kbit/s. The transceiver also protects the LAN from any faults that might occur in the station, usually by means of a *jabber* circuit. The jabber circuit ensures that if a transmitting station sends data packets that are too long, or it sends a continuous stream of bits, the station will be disconnected from the LAN. A bus LAN operates as a broadcast system; a station transmits its data to all the other stations but only the station to which the data is addressed is able to accept it.

Perhaps the main advantages of the bus LAN are that it is easy to add, change, or remove stations to/from the network without interrupting its operation and the network will continue to function if a node becomes faulty.

The main disadvantage is that the network is not able to cope adequately with heavy flows of traffic and its access protocol is complex. The main examples of bus LANs are Ethernet and token-passing.

Token-passing LAN

A token-passing LAN uses an access protocol that involves the continuous circulation of a 24-bit long unique data pattern *token* around a logical loop. The token is an access-granting message that circulates around the ring. Only one station can gain control of the token at a time. There is no central controller and all the stations are of equal status. A station with data to transmit must first gain control of the token. Access to the token is determined by the state of a bit at the leading edge of the token that indicates whether the token is free or is busy. If the token is free, a station with data to send will capture the token, transmit its data into the LAN complete with the address of both the sender and the destination station, and marks the token as busy. The token is then allowed to continue its passage around the LAN. The data is sent in the form of a number of packets; the dimensions of the loop are constant and so there is a fixed transmission delay around the logical loop. The loop can support a fixed number of bits at any one time and these bits are assembled into a number of separate packets. A station may transmit as many packets as it needs as long as the maximum time for which a station is allowed to retain the token is not exceeded. As the data circulates around the network its destination address is read by each station in turn until the destination station recognizes its own address; this station copies the data and then returns the (still full) packet to the LAN. When the packet has travelled right around the logical loop and has returned to the originating station, the data is removed from the packet and the token is marked as free before it is passed on to the next station.

The logical connection between the stations on the LAN is established by the passing of the token from one station to another. Each station has its own address, labelled as MA1, MA2, etc., in Fig. 10.2 and it is programmed with the address of the next station to which the token is to be passed. The 'next station' addresses are NA2, NA3, etc. In a ring LAN it is often the case that the token is passed from each station to the next station physically in the ring; then the labelling of the stations will go clockwise around the ring as shown. If some stations are to be given preferential service this is easily arranged by suitable NA addressing. If a station is non-operative passing the token to it is a waste of time and so the station passing the token will automatically pass the token to the next active station. This action bypasses the non-operative station and re-establishes the logical loop. The token is periodically offered to a non-operative station so that if it should become active again it will automatically become on-line. The idea is illustrated by Fig. 10.3 which shows a bus LAN in which station 4 is non-active; the logical loop is completed by station 3 having a next address of NA5 that will ensure that the token is passed from station 3 to station 5.

In many LANs some stations are only able to receive data and these stations are not offered the token. Each repeater that is serving a station that is able to both transmit and receive has two modes of operation,

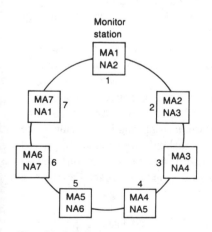

Fig. 10.2 Token-passing LAN.

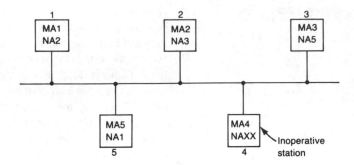

Fig. 10.3 Token-passing bus LAN.

known as the listen and the transmit modes. When a repeater is in the listen mode it checks the destination address of each packet to see if it matches the address of its associated station. When a match is found the data in the packet is copied and sent to the station, a bit in the packet is then changed to indicate that the data has been received, and then the packet is allowed to continue circulating around the loop. If the packet is not addressed for its station the repeater will retransmit the packet into the LAN. When a station has data to send the repeater must seize the token; it then enters the transmit mode and inverts the final bit in the token to effectively remove the token from the LAN. The station can then transmit its data via the repeater into the ring and this data is followed by the now marked busy token. When the transmitted packet arrives back at the originating station the data is removed from the packet, and then both the packet and the free token are sent on to the next station. A station that has just transmitted data cannot re-use the packet immediately but must wait for a short time before it can again seize the token. This procedure is adopted to prevent any one station monopolizing the system. Immediately a station has emptied a packet and sent on the token the station switches back to its listen mode.

The monitor station initializes the system and continually supervises the network and maintains traffic and fault statistics, etc. The format of a packet in a token-passing system is shown by Fig. 10.4. The two flags at each end of the packet each consist of the bits 01111110. This bit pattern is avoided in the message by the automatic insertion of a 0 after any sequence of five 1s.

Each station is connected to the Token Ring network by means of a *Trunk Access Unit* (TAU). Figure 10.5 shows how three TAUs may be employed to form a ring. The token-passing ring is very vulnerable to complete failure if one of its links should be broken. Improved reliability can be obtained by the use of the star-shaped ring shown in Fig. 10.6. If the link between any two centres is broken the separate

Fig. 10.4 Token-passing LAN packet format.

Eight-bit end flag	16-bit CRC	Message: up to 32 kbit	Eight-bit source address	Eight-bit destination address	Eight-bit token control	Eight-bit start flag

Direction of transmission →

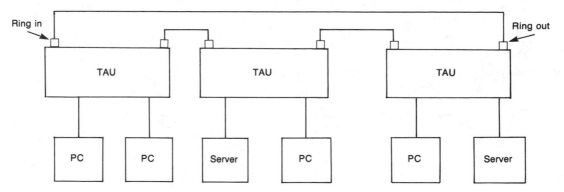

Fig. 10.5 Use of trunk access units (TAU) to form a ring.

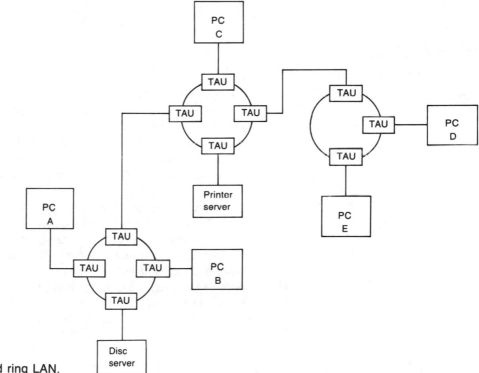

Fig. 10.6 Star-shaped ring LAN.

halves are able to continue working independently. Further reliability, at the expense of greater cost, is given by the dual ring network shown in Fig. 10.7. The dual cables would normally follow different routes between the computers so that if one cable becomes faulty the other would probably be unaffected.

Token-passing ring LANs are available in both 4 Mbits/s and 16 Mbit/s versions and do not suffer from the contention problems of Ethernet. However, the Token Ring system is more expensive to implement.

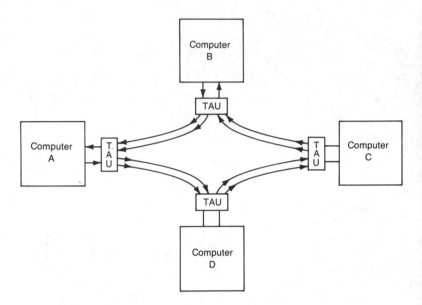

Fig. 10.7 Dual-ring LAN.

Ethernet LAN

The Ethernet LAN uses an access technique known as *carrier sense multiple-access with collision detection* (CSMA/CD). The technique allows any station on the network to attempt to gain control of the network at any time. The CSMA/CD control operates at each interface between the LAN and a station. Ethernet usually employs a baseband bus system in which the signals are modulated directly on to the cable. The cable is called the *ether* and each station is connected to it, via an interface cable, to a transceiver which is, in turn, connected to the LAN. The transceiver must be adjacent to the cable tap but the interface cable may be up to about 50 m in length. The basic arrangement of an Ethernet LAN is illustrated by Fig. 10.8 which shows a single *segment*. A segment may have up to 100 stations connected to it and it may be up to 500 m long. A segment is marked at 2.5 m intervals along its length and stations can only be connected to the LAN at these marked points. Each segment is terminated at

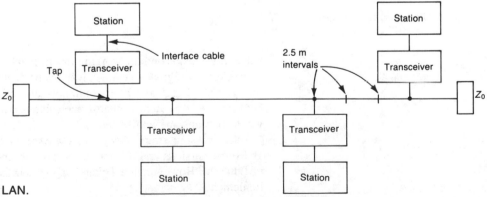

Fig. 10.8 Ethernet LAN.

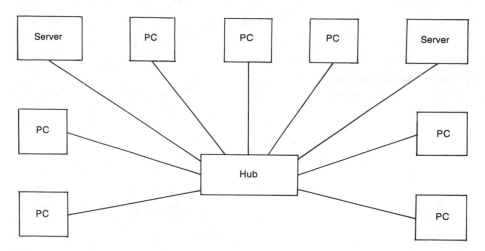

Fig. 10.9 Ethernet LAN using a star network.

each of its ends by an impedance equal to the characteristic impedance Z_0 of the cable.

The Ethernet stations are connected in a bus network using either thick or thin Ethernet coaxial cables. Thick Ethernet uses thick diameter copper cables which are terminated at each end by a *terminator* of resistance Z_0. Each PC, or other station, is connected to the cable by means of a transceiver and a drop cable. There may be up to 100 stations connected to one segment. The LAN can be extended by linking segments together using repeaters which regenerate the signal.

Thin Ethernet, also known as *cheapernet*, is considerably cheaper to install. It also uses a terminator at the end of each segment but it does not require the use of drop cable to connect a station to the LAN. This is because the software card in each station includes a transceiver.

There is also a third method of cabling that is sometimes used for Ethernet. It is known as 10base-T and employs unscreened twisted pair cables and an active hub-based star network. Figure 10.9 shows a typical arrangement. 10base-T is also used for fast Ethernet which is able to run at 100 Mbits/s.

If the hub in a star network is intelligent it will be able to mix both different Ethernet cable types and different LAN technologies, i.e. Ethernet, token ring and FDDI LANS can all be connected to the same hub. Some intelligent hubs support the *Simple Network Management Protocol* (SNMP), which allows traffic statistics to be collected.

An Ethernet LAN can be extended by connecting a number of segments together by means of repeaters; this is shown by Fig. 10.10. A repeater may be connected to a segment at any transceiver point and it transmits signals in both directions between two segments. A repeater incorporates protocols for dealing with possible data collisions and with any faults that may occur. If a link to a segment in another

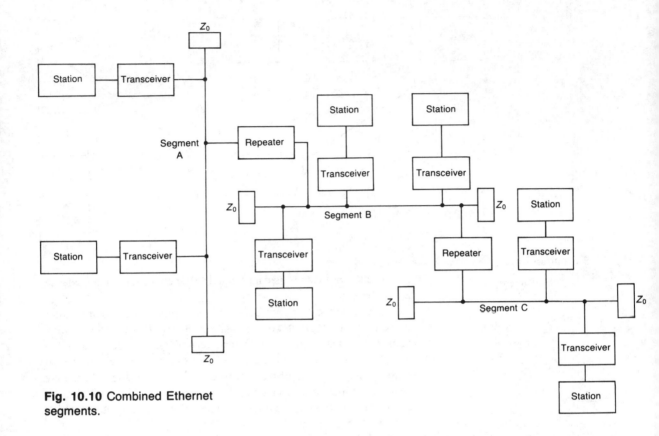

Fig. 10.10 Combined Ethernet segments.

building is wanted two remote repeaters, linked together by not more than 1 km of cable, will be employed. A fully segmented system may have up to 4000 stations in the network.

Ethernet does not employ a central controller and so all the stations connected to the LAN are of equal status. Any station on the LAN is able to try to transmit data over the LAN at any time; this is known as *multiple access*. Each station gains control of the network whenever it wishes to transmit data. CSMA/CD is used to deal with the simultaneous attempt by two, or more, stations to transmit data into the LAN. When a station has data to transmit it must first check to see whether the network is free or if it is already in use. If the network is busy the station must wait until it does becomes free. When the network is free the station can transmit its data into the LAN and this signal travels around the logical loop. All the other stations will then find the network to be busy if they should test it to find out. There is, however, an inevitable delay between a station commencing its transmission of data and *all* of the other stations becoming aware of this fact. During this short interval of time another station may also start to transmit data and then a *data collision* will take place. Each transmitting station compares its sent and received signals and if there is any difference between them a collision state is flagged. Immediately a transmitting station detects that a collision has taken place it stops

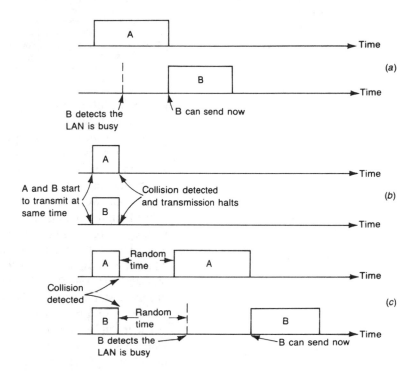

Fig. 10.11 Operation of an Ethernet LAN.

sending data and instead transmits a *jamming signal* to inform all the other stations of the collision. After a random interval of time each station with data to send will then make another attempt to retransmit and probably one of them will do so before the other(s) and so succeed in gaining control of the LAN. The principle of CSMA/CD is illustrated by Fig. 10.11. In Fig. 10.11(*a*) the LAN is initially clear and so the station A is able to transmit its data into the network. A short time later station B has some data to send but since it detects that the LAN is busy it waits until the data sent by station A has ended before it transmits its data. Figure 10.11(*b*) shows the situation when the LAN is initially clear and two stations, A and B, both transmit data at the same instant. A data collision is signalled to both the stations and their transmissions cease. Both stations then wait for a random length of time before they try again. Figure 10.11(*c*) supposes that station A retransmits before station B; station A's data is transmitted on to the network successfully this time but when station B tries again, after a slightly longer random time, it finds the LAN is busy. When station A's data transmission ends, station B recognizes that the LAN is now free and it is then able to send its data on to the LAN.

Each station may make up to 16 attempts to transmit data before abandoning the attempt and reporting an error. The number of data collisions will increase as the traffic on the LAN increases and although the system will work well when lightly loaded there may well be noticeable delays when the traffic is heavy. The maximum bit rate is 10 Mbits/s.

The format of an Ethernet signal is shown in Fig. 10.12. The 56-bit

Fig. 10.12 Ethernet LAN signal format.

32-bit CRC	368 to 12000 bit message field	16-bit length field	48-bit source address	48-bit destination address	Eight-bit start frame delimiter	56-bit preamble

preamble consists of the byte 01010101 repeated seven times; it is used for channel stabilization and synchronization. The preamble is followed by the eight-bit start frame delimiter which marks the start of the frame. Next comes the 48-bit address of the destination station and then the 48-bit address of the originating station; such a very large number of bits is used for each address because every Ethernet station in the whole world has its own unique address. In principle, therefore, every Ethernet station on the earth ought to be able to communicate with every other Ethernet station. The destination address may be either physical or multicast, and the first bit in the address specifies which it is. If a multicast address is given it denotes a multi-destination message; for example, a multicast group address may refer to a group of related stations while a broadcast address refers to all the stations connected to a particular Ethernet LAN.

The length field indicates the number of bytes in the message field which may contain anywhere in between 368 and 12 000 bits. Lastly the 32-bit CRC uses the generating polynomial $X^{32} + X^{26} + X^{23} + X^{22} + X^{16} + X^{12} + X^{11} + X^{10} + X^8 + X^7 + X^5 + X^4 + X^2 + X + 1$.

Compared with a ring system a bus system, such as Ethernet, has the advantage of fully distributed network control which makes it both cheaper and simpler to operate. Also the failure of one station does not affect the whole network.

When fibre optic cable is used as the transmission medium an Ethernet LAN may use the star configuration and an example of this is shown by Fig. 10.13. The hub is able to detect data collisions as well as distributing packets to all ports except the source port. The multi-port repeater can also perform these functions as well as the reconstruction of preambles and the regeneration of timing pulses.

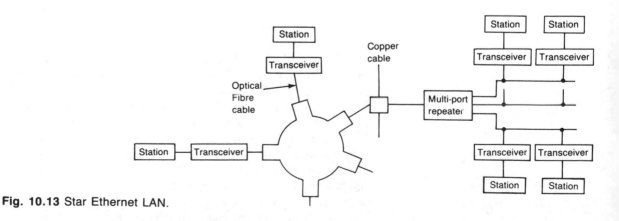

Fig. 10.13 Star Ethernet LAN.

Fibre Distributed Data Interchange

Fibre Distributed Data Interchange (FDDI) uses optical fibre cable to form a dual token-passing ring that passes packets of data and is able to operate at bit rates as high as 100 Mbits/s over distances of up to 100 km with up to 500 nodes that are anything up to 2 km apart. The dual rings rotate in opposite directions and one of them is intended to be a redundant ring that is only used if the other ring becomes faulty. FDDI is more expensive to implement than Ethernet or token passing over copper cable. Because of this FDDI is usually employed to provide a backbone network linking LANs, and not as a means of linking PCs and peripherals.

A development of FDDI has led to the use of screened twisted pair copper cable, and although this is cheaper than optical fibre FDDI it is still more expensive to implement than Ethernet and token ring LANs.

Servers

Hardware known as *servers* is often employed in LANs to increase the number of stations that can be connected to the LAN and/or have simultaneous access to common files. Commonly employed are disk servers, file servers, printer servers and terminal servers.

Disk Server

A disk server is used to provide access to a hard disk store for a number of small computers; the server is transparent to the user so that each user is under the impression that he is accessing a local disk drive. The files and programs that are stored on the hard disk can be used by each station given access just as though the files and programs were stored on a local disk drive. Figure 10.14 shows an example; a number of PCs, each with its own disk drive, are connected to the LAN and thence to the disk server. When one of the computers wishes to access the hard disk store it sends a request to the disk server and this then provides the desired access. Any operational differences such as different bit rates and frame formats between the hard disk memory and the accessing station can be resolved by the server since it is able to provide speed, etc., translation. The use of a disk server allows each of the PCs to have the use of an expensive, large-capacity,

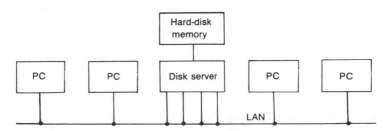

Fig. 10.14 Use of a disk server.

hard disk memory which otherwise would probably be uneconomic to provide.

File Server

A file server provides a service similar to that of the disk server but it also manages the local disk drives at the individual computers. A file server is based upon disk-held software that manages the stored files and allows some, or perhaps all, of the stored data to be shared between different users. The idea is not restricted to files; printers and other output devices can also be accessed as though they were files.

Terminal Server

A terminal server acts like a kind of multiplexer to allow several small computers, or other terminals, to have access to the same node on a LAN. An example of the use of a terminal server is shown by Fig. 10.15. A terminal server may be used to provide low-cost access to a host computer for a number of terminals. For PCs, running terminal emulation software, the terminal server provides flexible communication to a host computer. It also allows a group of PCs to share access to various peripherals.

At the simplest level a terminal server merely provides compatible inter-connection between EIA 232E devices, at their most complex terminal servers form the basis of a terminal network for a mainframe

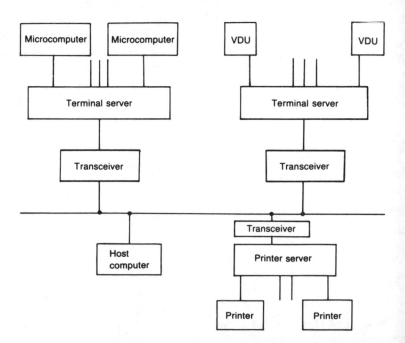

Fig. 10.15 Use of a terminal server.

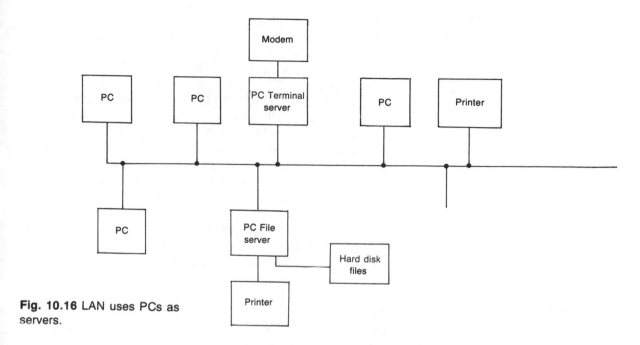

Fig. 10.16 LAN uses PCs as servers.

computer. A terminal server is able to give up to 8 EIA 232E peripherals access to a LAN via a single connection. Any differences between bit rates or frame formats can be sorted out by a server, and data flow control is provided.

In many LANs the servers are themselves PCs. A PC that is to act as a dedicated server must be powerful enough to cope with the anticipated workload and should run at a fairly fast speed, probably at least 50 MHz. A file server will also require a high-capacity hard disk with a fast access time. Usually a file server will have more than one hard disk drive. The interface between a server and the LAN is made by means of a network adaptor. If a PC is fitted with file server software, other PCs will be able to use its hard disk just as though it were another hard disk on their own PC. Printer server software must be supplied to the PC that is to act as the printer server. As a file is printed out the printer software will put any other printing requests into a queue and print them in sequence: this is known as printer spooling. The use of a printer server with an expensive printer does not stop other PCs having their own printer (almost certainly a cheaper type) that can be used when required.

An example of a LAN that uses PCs as servers is shown by Fig. 10.16.

Peer-to-Peer Networks

A LAN with up to about 10 PCs may be operated on a *peer-to-peer* basis, in which each PC has equal status. Each PC is able to access all the peripherals attached to the LAN, such as printers and disk drives, and all the other PCs are able to use any peripherals connected to that PC. This means that each PC must be supplied with software that allows it to act as a non-dedicated server. For example, each PC will be able to use the hard disk connected to another machine just as though it were a second hard disk connected to itself.

Because each PC has to act as both a PC to its local user and as a server to the user of any other PC, some contention is inevitable. If the local user of a PC is carrying out some work that uses most of the capacity of that PC, a remote user may well experience some delay before his server request is answered. Conversely, if a remote user is accessing a large file on a PC via the server software, the local user may find that his PC's performance is sadly lacking. To obtain the best performance from a peer-to-peer LAN, the storage of files that are used by many people should be spread over a number of PCs.

A peer-to-peer network is intended for light usage and is restricted to small-scale LANs that have a limited number of nodes. Most peer-to-peer LANs use the Ethernet system.

Inter-networking

The usefulness of a LAN is usually greatly increased if its users are given transparent access to one, or more, other LANS and/or to a WAN. The interconnection of LANs is known as *inter-networking*. Inter-networking gives a user access to additional facilities, additional data, and also, of course, to a greater number of other stations. Frequently the LANs which are linked together employ incompatible protocols but they must always use the same language.

Bridges

Circuits known as *bridges* are employed to connect two LANs together and allow packets of data to travel from one LAN to another. A bridge provides a link between two similar types of LAN, e.g. two Ethernet or two token-passing LANs. It extends a LAN over a wider area, making all the inter-connected segments appear to be part of one much larger LAN.

A bridge will only transmit those data packets from one segment to another which have been addressed to a server, or a PC, in the other segment. To allow a bridge to filter the traffic between two LANs the bridge is supplied with an internal table of network addresses, which contains the address of every node on both sides of the bridge. When a data packet arrives at the bridge its destination address is checked against the address table. If the destination address is on the local LAN the bridge will ignore the packet, but if it is not a local address the packet will be passed through to the other LAN. This action may be repeated over several different LANs until the destination LAN is reached.

When a Kilostream circuit is used to connect two bridges each bridge is connected directly to the NTU.

Routers

A router routes packets of data between different LANs as well as undertaking various network management tasks such as network traffic control. A router must be able to integrate differing LAN protocols, NetWare and DECnet for example, and interface to a host computer. The usual way in which this is done is to run a third protocol on both the linked LANs concurrently with the existing (but incompatible) protocols. There is only one protocol that is currently able to do this and it is known as *Transmission Control Protocol/Internet Protocol* (TCP/IP). TCP/IP is supported by almost all kinds of computers from PCs to mainframes, and it provides both terminal emulation and file transfer facilities. A typical router network using TCP/IP is shown by Fig. 10.17.

Figure 10.18 shows an example of two token-passing LANs connected both together and also to an Ethernet LAN by a router. Two, or more, routers can be connected together by WAN links, such as Kilostream and Megastream, and Fig. 10.19 gives an example of such a network.

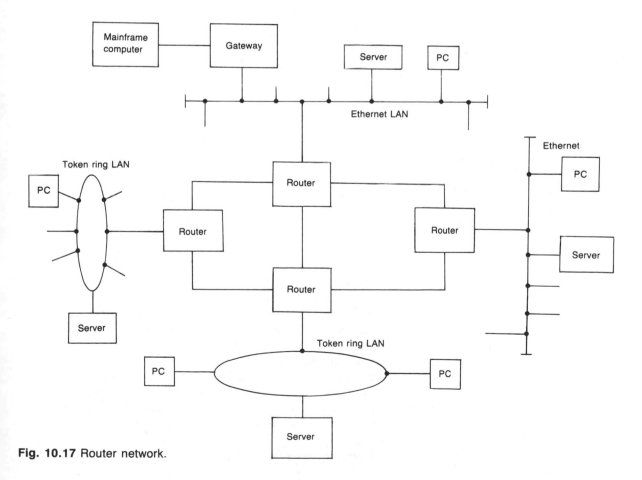

Fig. 10.17 Router network.

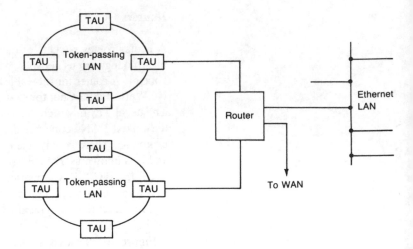

Fig. 10.18 Use of a router to interconnect LANs.

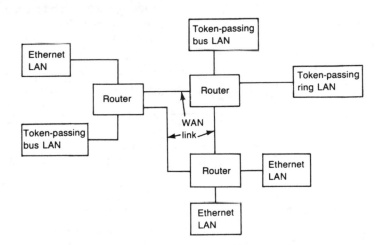

Fig. 10.19 Multi-LAN network.

A router normally has access to a routing table which holds information about the network. The router can use this information to select a path through the network to link a station in one LAN to its required destination in another LAN. When a data packet is received the router looks up the internal routing table to determine possible routes to the destination LAN. If, for example, a PC connected to LAN A in Fig. 10.20 wishes to communicate with a database connected via a server to LAN C there will be three possible routes, any one of which might be selected. Route 1 is direct from LAN A to LAN C, route 2 is via LAN B, and route 3 is via LAN D. The router can then select the best route to be used; *best* depends upon such factors as speed, cost and availability. If a link should become faulty the router should be able to choose an alternative routing.

Most routers support the *Internet Protocol* (IP), which is a subset

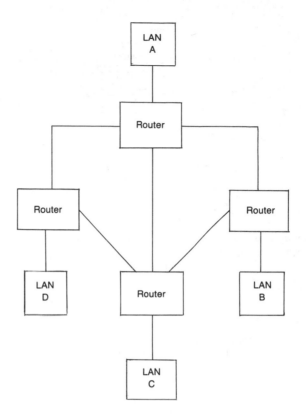

Fig. 10.20 Router finds alternative routes between LAN A and LAN C.

of the *Transmission Control Protocol/Internet Protocol* (TCP/IP) and/or *Internet Packet Exchange* (IPX). These protocols make it possible for servers using different software to be inter-connected; the router must be able to determine the protocol of each data packet and then decode its addressing information.

TCP/IP has become an industry standard for host computer inter-networking. It allows full intercommunication between all nodes both within a single LAN, between separate LANs, and between LANs and WANs. The TCP/IP family includes protocols for (*a*) electronic mail, (*b*) file transfer, (*c*) printing, (*d*) remote command execution, (*e*) remote login, and (*f*) network maintenance. Each protocol has two forms: user modules that request a service and server modules that provide a service. TCP/IP can be run on most PC LANs. A *Brouter* is a combination of a bridge and a router.

Gateways

The most complex form of connection between two LANs is known as a *gateway*. A gateway is used to inter-connect two LANs that employ different protocols, such as a PC network and a mainframe computer. Often a gateway is a PC supplied with controller emulation

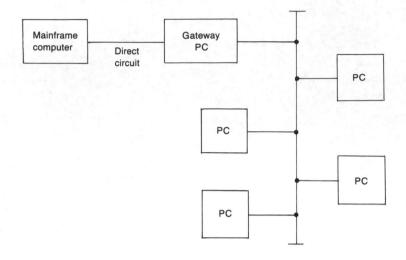

Fig. 10.21 Gateway inter-connects a mainframe computer to a LAN.

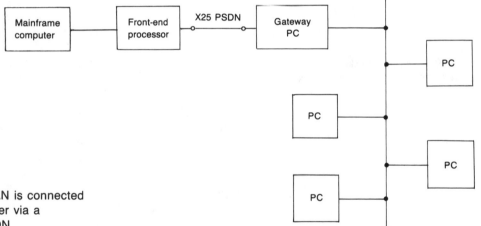

Fig. 10.22 Distant LAN is connected to mainframe computer via a gateway and the PSDN.

software. Figure 10.21 shows how a gateway may be used to link a mainframe computer to a LAN. The gateway runs software that emulates a cluster controller to enable a number of PCs to be given access to the mainframe. The PCs must each be fitted with terminal emulation software. The gateway handles the communications protocols on behalf of the PCs. Sometimes, as shown by Fig. 10.22, connection to the mainframe may be via the PSDN.

Terminal Emulation

Most mainframe (host) computers require terminals to act in a specified manner with regard to such things as their response to commands to (*a*) update the screen, (*b*) position the cursor, and (*c*) read the keyboard. Some host operating systems, such as UNIX, are easily configured to support a wide range of terminals. Terminal emulation software allows a PC to act as though it were a mainframe terminal.

Broadband Local Area Networks

A broadband local area network uses frequency division multiplex techniques to divide the bandwidth of a coaxial cable into a number of 6 MHz channels. The channels are then divided into two groups: the higher-frequency channels are used for transmission from the head end transponder to the remote stations while the lower-frequency channels are used for transmissions in the opposite direction. The basic arrangement of a broadband LAN is shown by Fig. 10.23. The head-end transponder is connected to the main coaxial cable into which a number of taps and amplifiers are inserted. The taps are wanted at each point where some stations are to be connected to the line and an amplifier is used at regular intervals along the line to keep the signal strength high. Where required the main line can be split into two halves by means of a circuit known as a splitter. At each tap a drop coaxial cable is used to connect the tap to a LAN controller, this consists of a radio-frequency modem and access logic. The access logic connects the LAN controller to one, or more, stations via V24 interfaces.

A station that has data to transmit to another station sends a request to the system controller that is located at the head end of the system. This then assigns suitable frequencies to the calling and the called

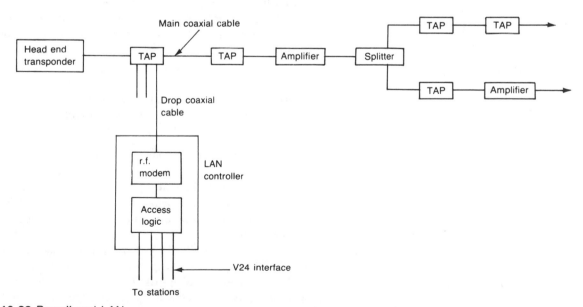

Fig. 10.23 Broadband LAN.

modems and notifies them of these frequencies. Each r.f. modem must then tune itself to its allotted frequency. The transmitted data is then applied to the modem where it is converted to modulation of the allotted carrier frequency and then sent to the head-end transponder. Here the signal is frequency shifted to the frequency band allocated to the receiving r.f. modem. Since a modem may be assigned any frequency within a given frequency band, all modems must be capable of rapidly changing their transmitting and receiving frequencies, i.e. the modems must be *frequency agile*. This method of operation has some similarities to a dial-up connection via the PSTN. If two stations have a considerable amount of traffic between them they can be permanently assigned a particular carrier frequency and have both their modems tuned to that frequency: this operation is analogous to the use of a dedicated leased line between two terminals. Another method of operation that is often employed is to allow all modems to operate at the same frequency in each direction of transmission. Some form of access protocol is then necessary such as token passing or CSMA/CD.

NetStream

NetStream is BT's own family of LANs which has four members.

(a) NetStream Baseband is an Ethernet LAN that uses CSMA/CD at 10 Mbit/s.

(b) Netstream Broadband is a system that combines multiple data and video networks and can be used for the in-house distribution of high-speed data and video signals.

(c) NetStream Gateway integrates NetStream Baseband directly with the BT's Switchstream service (p. 183).

(d) NetStream Fibre Optic allows optical-fibre cable to be used in a NetStream Baseband system to link together two coaxial cable segments.

IEEE 802 Specifications

A set of standards for LANs has been produced by the USA Institution of Electrical and Electronic Engineers (IEEE). Briefly these standards are as follows:

(a) 802.1 is an introduction to the 802 family and it is concerned with the relationships within the family and with their relationship with the OSI model (p. 201).

(b) 802.2 describes the functions and protocols of the logical link control that is common to all media. It describes the peer-to-peer protocol.

(c) 802.3 defines the CSMA/CD bus access system.

(d) 802.4 defines the token-passing bus access method.

(e) 802.5 defines the token-passing ring-access method.

(f) 802.6 deals with metropolitan area networks (MANs) that

provide data, speech and video communications between buildings within a relatively small geographical area such as a city.

Many of the IEEE 802 standards have been adopted by the ISO as 8802 standards.

A metropolitan area network (MAN) extends the idea of a LAN into a network that could cover a city with a diameter of about 50 km. The transmission medium is generally fibre optic cable and any station on the network is able to communicate with any other station. Messages transmitted by one station to another are sent in packets with the address of the destination placed in the header. Neither Ethernet nor token passing will work properly over the distances involved and so another system, known as *distributed queue dual bus* (DQDB), is used instead.

Cableless Local Area Network

A *cableless*, or *cordless local area network* (CLAN) is the combination of a LAN and mobile radio, and a version of the star network is shown by Fig. 10.24. The stations are connected to the hub by radio links and no expensive cabling is required. Each station is supplied with a network interface card and an aerial. Alternative arrangements would include the equivalents of the ring and bus networks. The radio frequencies used are either 2.445 to 2.475 GHz or 18 GHz and the bit rate is about 2 Mbits/s. The most common application of CLANs is in supermarkets and department stores using PoS terminals.

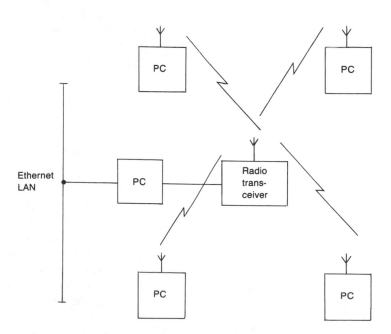

Fig. 10.24 Cableless LAN.

11 Message Switching and Packet Switching

Any large data network has many computers, such as mainframe, mini-, and PC, and data terminals of various kinds connected to it. Any of these terminals* may require access to any other terminal and so some form of switching is necessary. Synchronous data can be switched through a network using one, or more, of three different techniques; namely, *circuit switching, message switching* or *packet switching*. Circuit switching has already been discussed, (in Ch. 6), and it involves the setting up of an end-to-end transmission path through a network for the duration of a call, whether or not either terminal is transmitting data. This means that the maximum possible utilization of each link is not achieved. When an organization depends upon the assured delivery of messages within a fairly short time it cannot afford to rely upon the vagaries of data transmission via the PSTN. Switched circuits via the PSTN have the following disadvantages.

(a) The dialled-up connection must be retained for the duration of the call even though periods when no data is transmitted may occur.

(b) Both terminals must be operative at the same time in order to be able to communicate.

(c) Both terminals must operate at the same transmission speed.

Some of these disadvantages are overcome by the use of dedicated circuits and/or a circuit switched private system but this may not always be an economic answer. Two alternatives are the use of either message switching or packet switching. In a message-switching system a message is transmitted by a terminal to its nearest switching centre without any need to wait until a link has been set up with the destination terminal. The message is stored at this switching centre until it can

*The term 'terminal' is taken to apply to both a computer and a terminal.

be forwarded to another switching centre that is nearer to the destination terminal. Here the message is again stored until it can be passed on to the next switching centre that is even nearer to its destination, and so on. Eventually, the message will arrive at the nearest switching centre to the destination terminal; here it will be stored until the terminal is ready to accept the message.

The third method of switching is known as packet switching. In a packet switching system the data traffic is divided up into a number of packets, each of which is given a label to identify it with a particular connection through the network. A connection only uses bandwidth when it is carrying packets, otherwise the bandwidth is available for other users. Terminals are connected, either directly or via a PAD, to the nearest packet-switching exchange (PSE). Before a message is transmitted through the packet-switching system it must first be divided into a number of short blocks or *packets*. Each packet contains enough addressing information for it to be transmitted through the network to its destination. Here the packets are reassembled to obtain the original message; the reassembly is carried out either at the destination terminal or at the distant PAD. In the UK a public switched data network (PSDN) is provided by BT under the name of Packet Switchstream (PSS) and through International Packet Switchstream (IPSS) users have access to the PSDNs of several other countries. Some private packet switching systems also exist, either as a WAN or as an alternative to a LAN, and these networks are often provided with access to the PSS.

Message Switching

When a terminal has a message to send to another terminal it can immediately transmit it into the message-switching system; there is no need to either set up a link to the destination terminal or for the destination terminal to be ready to receive data. Nor does it matter if the two terminals operate at different transmission speeds; the system will automatically convert the message to the speed of the receiving terminal. The message is transmitted into the system with an added header that includes the addresses of both the source and the destination terminals and some control information. The system will check the message for errors and, if necessary, request retransmission; the error may be in the source or destination addresses or in the format of the message as well as in the message itself. The message will be stored at each switching centre while the centre determines a suitable route to the next switching centre and a channel over this route becomes available. The message is then taken from the store and forwarded to that centre. Usually a centre will have a choice of routes. A switching centre must have received and stored the complete message before it can forward it to the next stage in the link. This is known as a store-and-forward system. The message is retained in the store after it has been passed on to the next switching centre in case there is an error in its transmission. This procedure is repeated as many times as is necessary until the message arrives at the switching centre nearest to its destination terminal. Here it will be passed to the terminal unless the terminal is either busy or it is non-operative; if so the message will be held in store, perhaps for several hours or even days, until it can be delivered. Alternatively, the system may have been notified of an alternative terminal that can accept the message and in this case it will reroute the message to this other terminal.

As soon as a message has been passed over a channel another message, most likely being sent between two completely different terminals, can pass over the channel. This gives a very high utilization of each channel in a message-switching network. The basic concept of a message-switching centre is shown by Fig. 11.1. A message M is to be sent from the terminal T1, connected to switching centre SC1, to computer C2, connected to switching centre SC4. The message is first passed from T1 to SC1 and here it is stored until a free channel to either SC2 or SC3 becomes available. In the figure it has been assumed that a channel to SC2 becomes available first and so the message is forwarded to SC2. At switching centre SC2 the message is again held in store until a free channel to SC4 becomes available. Once a suitable channel has been obtained the message is passed on to SC4 and here it is stored until computer C2 is able to accept it.

At each switching centre the stored messages are placed into a queue waiting for transmission to the next stage in their route through the network; this may introduce considerable delays at times of heavy traffic. Some messages can be given a priority rating which enables them to jump the queue, also a message can be given a multiple destination address so that it will be broadcast to several terminals.

The advantages claimed for message switching are as follows.

(a) Messages can be transmitted at any time convenient to the sender without the need to check to see if the recipient is busy or even operational.

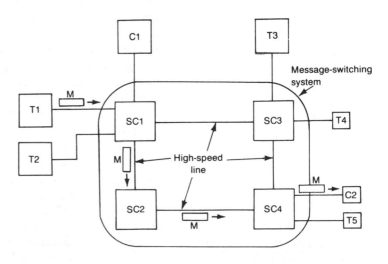

Fig. 11.1 Message-switching network.

(b) The network automatically performs code, protocol, and speed conversions and this permits different types of terminal to communicate with one another.

(c) The queueing of messages and automatic dialling gives a high utilization of the lines.

(d) Messages can be broadcast to several terminals.

(e) If the traffic flow is heavy calls are not blocked but merely delayed.

Packet Switching

A packet-switched network consists of a number of packet-switching exchanges (PSE) that are interconnected by time-division multiplexed high-speed lines. Each user of the network can have a wide variety of terminals, including dumb terminals, front end processors, routers, and PADs, and is connected to the nearest PSE by a *dataline*. Each message that is transmitted into the packet-switched network must first be divided into a number of packets before it enters the network. If the terminal is intelligent, such as a PC, when it is known as a *packet terminal*, it will be able to perform this packet assembly and it will be connected directly to the network. A non-intelligent terminal, known as a *character terminal*, is unable to convert messages into packets and so it must be connected to the network via a *packet assembler/disassembler* or PAD. A PAD permits terminals that do not have a suitable interface for direct connection to a PSS to obtain access to such a network. Figure 11.2 shows a packet-switched network that has four packet-switching exchanges to which are connected a number of computers, intelligent terminals and non-intelligent terminals. The packet-switching exchanges are interconnected by high-speed Megastream circuits.

Each message transmitted into the packet-switched network must previously have been divided into a number of packets. Each packet

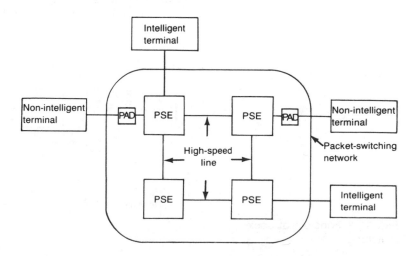

Fig. 11.2 Packet-switching network.

has a specified format and may be up to 1024 bits in length, and it is transmitted separately through the network. When a packet reaches the local PSE it is enveloped with a header and a trailer, (see page 190). All the packets may follow the same route through the network or they each may take different routes. The former method of routing is now the more common technique; it is known as a *virtual call* and it has been standardized by the ITU-T X25 recommendations. The term *call* means an exchange of data between two terminals via a packet-switched network and a packet is an individual part of the data carried by a call. As each packet arrives at a PSE its destination address is examined and if the destination terminal is connected to this PSE the packet will be passed immediately on to the terminal (unless it is either busy or non-operative). If the packet is to be passed on to another PSE it is placed in a queue — unless it has a priority rating — along with the packets from other terminals. When a packet reaches the top of the queue it will be forwarded over the first suitable channel to become available. When all of the packets making up a particular message have arrived at the final PSE in a link the packets must be disassembled to obtain the message. This disassembly is carried out either at the terminal (if it is intelligent) or by a PAD.

The basic principle of packet switching is shown by Fig. 11.3. Packet terminal 1 has a message to transmit to packet terminal 4, and packet terminal 2 has a shorter message to send to packet terminal 3. The message transmitted by packet terminal 1 is split up into three packets P_1, P_2, and P_3 before it is transmitted to the nearest packet-switching exchange PSE 1. At this exchange the packets may be switched to any link that is available and that provides a possible route to the destination PSE. In the figure the packets are switched to the link that connects with PSE 2. At the same time the message sent by packet terminal 2 has been divided into two packets, P_4, and P_5. When a free channel to packet switching exchange PSE 4 becomes available all five packets are transmitted towards PSE 4 using time-division multiplex (TDM). At PSE 4 the packets P_1, P_2, and P_3 are passed to their destination, packet terminal T4, while packets P_4 and

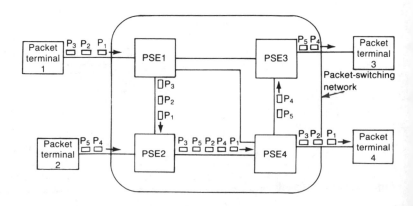

Fig. 11.3 Principle of packet switching.

P_5 are switched on to a channel of the route to PSE 3 and thence to their destination.

ITU-T X Recommendations

The ITU-T have produced a number of network interface standards for packet-switched networks which specify the protocols for the reliable transfer of data across the network. These are given in Table 11.1.

For messages to be interchanged successfully between a terminal and a packet switching network a standard network-access protocol is necessary. This protocol is provided by the ITU-T X25 recommendation. The protocol provides the following facilities to the user.

(*a*) It divides the message into packets.
(*b*) Each packet is checked for error and if an error is detected the packet is retransmitted.
(*c*) An international format for destination addressing is given.

Table 11.1 ITU-T Packet-switching standards.

X.1	International user classes of service and data signalling rates for a PSDN.
X.2	International user facilities available in a PSDN.
X.3	Packet assembly/disassembly facility (PAD) in a PSDN.
X.4	General structure of signals for data transmission over a PSDN.
X.20	Interface between a DTE and a DCE for non-synchronous transmission on a PSDN.
X.20bis	V21 compatible interface between a DTE and a DCE for non-synchronous services on a PSDN.
X.21	General-purpose interface between a DTE and a DCE for synchronous users in a PSDN.
X.21bis	Use of a DTE on a PSDN that are designed for interfacing to synchronous V-series modems. It is a refinement of V24.
X.24	List of definitions of interchange circuits between a DTE and a DCE for terminals operating on a PSDN.
X.25	Interface between a DTE and a DCE for terminals operating on a PSDN.
X.26	Electrical characteristics for unbalanced double-current interface circuits for general use with IC equipment.
X.27	Electrical characteristics for balanced double-current interchange circuits for general use with IC equipment.
X.28	DTE/DCE interface for a non-synchronous DTE accessing the PAD in a PSDN.
X.29	Procedures for the exchange of control information and user data between a packet mode DTE and a PAD.
X.75	Interface for interconnecting PSDN.
X.92	Hypothetical reference connections for PSDN.
X.95	Network parameters for PSDN.
X.96	Call progress signals in a PSDN.
X.121	International addressing scheme for PSDN.

Note: DTE = data terminal equipment such as computers and terminals of all kinds. DCE = data circuit-terminating equipment, i.e. packet-switching equipment.

(*d*) The user is unaware of any differences in the speeds of the calling and the called terminals.

(*e*) Control of the data flow through the network is given to the PSEs.

Packet terminals are able to implement the ITU-T X25 recommendations and so such terminals may be connected directly to a packet switched network. Character terminals, however, are unable to implement X25 and so they must be connected to a packet-switched network via a packet assembler/disassembler, or PAD. A PAD accepts data from a character terminal, executes the X25 protocol, and then passes the resulting packets to the packet-switching system. The PAD does not alter the contents of the message in any way, it simply assembles the message into packets in one direction, disassembles the packets into the original message in the other direction, and handles call set-up, addressing and flow control. Most PADs have more than one port and this makes it possible for them to be shared between a number of character terminals.

Per-packet and Per-call Routeing

Changes in traffic flow, and line and/or equipment faults, mean that the best route between two terminals may vary with time. The use of fixed routes through the network is therefore inefficient and so adaptive routeing is generally employed. The movement of packets through a packet-switched network can be achieved in two ways: (*a*) per-packet routeing, and (*b*) per-call routeing. Per-call routeing is the more commonly employed method and its use is specified by the ITU-T X25 recommendations.

Per-packet Routeing

With the per-packet routeing method of operation each packet of data is given the address of its destination terminal and it is treated by each PSE as a distinct and separate entity. Although all of the packets are delivered to the same destination they arc each given an individual routeing through the network. As the packets carrying a certain message arrive at the first PSE a routeing decision will be made for each packet, this will be based upon the queue for each link to other PSEs. Each packet will be routed over the path to the next PSE that, at that instant, will give the best performance. This allows temporarily congested, or failed, links to be avoided. Often this will result in the packets of a particular message following different routes through the network. This may well result in the packets arriving at their destination in the wrong order; they will then have to be re-sorted into the correct sequence before their disassembly into the message is carried out. The sorting out, and the disassembly, will be carried out by either the destination terminal or a PAD as will the recovery of any lost or corrupted packets and the rejection of any duplicated packets. This kind of routeing is not used in the PSDN but it is used in some private networks.

Per-call Routeing

For the per-call method of routeing packets through a network a path is selected for each terminal-to-terminal communication and this *virtual circuit* is set up before the packet transmission begins. A virtual circuit is a logical point-to-point connection between a sending terminal and a receiving terminal. The virtual circuit is not a dedicated path as in circuit switching, but a label that identifies a route across the network. Once a virtual circuit has been set up its number replaces the destination address in the header. This reduces the overhead because the number takes up less space than does the address. To route a packet across a node the receiving node must extract the packet from its frame, read the number of the virtual circuit, identify the wanted output node, put the packet into a new frame, and then the packet can be sent on to the next node. Each packet is stored at each PSE and put in a queue for an output link to the next stage in the chosen route. The packet-switching network routes all the packets associated with a particular message over the same path through the network and so they will arrive at their destination in the correct order. The network is responsible for the recovery of any lost or corrupted packets and also for the non-delivery of any duplicated packets. The virtual-circuit method is specified by the ITU-T X25 recommendations; the protocol defines the procedures for the setting up of a virtual circuit, the transfer of packets over the virtual circuit, and for clearing the virtual circuit once all of the data has been sent. If two terminals often have large amounts of data to transfer between them they may be provided with a *permanent virtual circuit* (PVC).

X25 has three levels. Level 1 is concerned with the interface between the terminal and the packet-switching network. Level 2 gives the procedures for the setting up and the maintenance of a link between the terminal and the network; at this level the HDLC protocol, or LAP-B, is implemented. Level 3 specifies the manner in which a message is formed into packets and it also gives the rules for the establishment and control of the flow of data.

Packet Format

A data message is divided into short blocks whose maximum length is 4096 bits and it is then placed into one, or more, packets. Each packet has a header that contains routeing information which is used to switch the packet through the network to its correct destination terminal. The packet is also given a frame header and a frame trailer which provide an HDLC framing envelope. The format of a packet

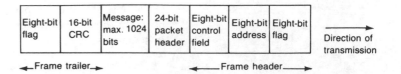

Fig. 11.4 Packet format.

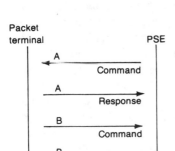

Fig. 11.5 Use of an eight-bit address field.

is shown by Fig. 11.4. The packet header follows the X25 recommendations and the frame header and trailer follow the HDLC protocol or a subset of it such as LAP or LAP-B.

The two start flags define the beginning and the end of the frame within which the error detection process is applied. The flag sequence is 01111110 and it must not be allowed to appear anywhere else inside the packet. As each terminal transmits a message it monitors the transmitted packets and if five consecutive 1s should be detected, other than in a flag, it will immediately bit stuff a 0. At the receiving terminal the incoming packets are again monitored and whenever five 1s followed by a 0 arrive the 0 will be discarded or bit stripped. The eight-bit address field does *not* give the address of either the source terminal or the destination terminal (this is given in the packet header). Instead, the 'address' is a code that is used to indicate whether a frame is one that just been originated — known as a command frame — or whether it is an acknowledgement — known as a response frame. This makes it possible for the network to distinguish between a frame that contains data that must be passed on to another PSE, or to a terminal, and a frame that can be discarded. There are just two possible 'addresses': 00000011 is address A and 00000001 is address B, and Fig. 11.5 shows how they are used.

Each bit in the eight-bit control field has a separate function. The more important of these are the following.

(a) The first bit transmitted is set to 0 to indicate an information frame, and is set to 1 to indicate either a supervisory frame (i.e. one used in the setting up of a virtual circuit) or an un-numbered frame (i.e. one concerned with the acknowledgements involved with setting-up and with error control). There are three kinds of supervisory frame: the receive ready frame (RR), the receive not ready frame (RNR), and the reject frame (REJ).

(b) If a frame is transmitted but is then not acknowledged it will, after a certain time delay, be retransmitted. The fifth bit is the poll/final bit and it is normally set at 0. When it is at 1 either the retransmission of the packet is requested, or the packet is one that has been retransmitted. Bits 2, 3 and 4 indicate, using binary arithmetic, the number of a frame (0 to 7), and bits 6, 7 and 8 indicate the number of the next frame that is expected in the opposite direction.

This counting provides the means by which it can be ensured that frames, and hence packets, are sent to the destination terminal in the correct order. Lastly, the CRC operates to detect any errors in the manner described in Chapter 9. It works by storing a copy of each transmitted packet at the sending node until an acknowledgement of error-free receipt is received from the receiving node. If acknowledgement is not obtained the packet is continually re-transmitted until an acknowledgement is received. The error-checking feature also provides flow control; if a receiving node is becoming overloaded it will delay sending an acknowledgement, even though receipt was error free. This delay will then reduce the rate at which the sending node transmits new packets.

Example 11.1

Write down the contents of the control field if the control information being passed by PSE A to PSE B is: (*a*) This is frame 1 from A; frame 2 from B is wanted. (*b*) This is frame number 4 from A; frame number 6 from B is wanted.

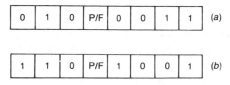

(*a*)

(*b*)

Fig. 11.6

Solution
The required control fields are shown in Figs. 11.6(*a*) and (*b*), respectively.

Terminal-to-PSS Link Set-up

There are two methods that may be used for the setting-up of the link between a terminal and the nearest PSE; the two methods are known as the link access procedure (LAP) and the link access procedure-balanced (LAP-B). If LAP-B is used the end that wishes to set up the link must first transmit the command SABM and this must be acknowledged by the other end before any data can be sent. If LAP is used the command is SARM; this must first be acknowledged by the other end, then repeated and acknowledged in the opposite direction. The end that originates the setting-up of the link is known as the primary station and the other end is then the secondary station. The setting-up of a link, and its later clearing down, follows the procedure shown by Fig. 11.7. The PSE sends the command DISC(P) to the called terminal to inform it that a call is coming and the terminal acknowledges it by returning UA(F). The terminal then sends SABM to establish the LAP-B link protocol and this is acknowledged by the PSE sending back UA. A number of information frames are then transmitted in both directions until the final frame is signalled by I(P). The receipt of this frame is acknowledged by the PSE returning RR(F). The terminal then sends the DISC command to clear down the link and this is acknowledged by the PSE returning UA.

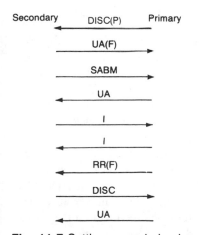

Fig. 11.7 Setting-up and clearing-down a terminal-to-PSE link. DISC(P) = link required; UA(F) = response to DISC; SABM = LAP-B link set-up command; I = information frame; DISC = link clear-down command.

PSE-To-PSE Link Set-up

A packet-switching network may use two types of call that are known respectively as the data call and the fast select call. Both types of call

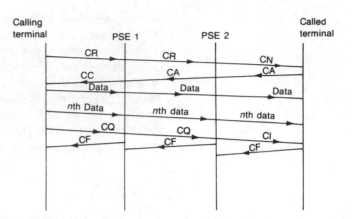

Fig. 11.8 X25 setting-up procedure.
CR = call request (00001011);
CN = incoming call (00001011);
CA = call accepted (00001111);
CC = call connected (00001111);
CQ = clear request (00010011);
CI = clear indication (00010011);
CF = clear confirmation (00010111).

can be used by packet terminals but a character terminal can only make data calls. A data call has three parts: (*a*) call set-up, (*b*) data transfer, and (*c*) call clear down. Figure 11.8 shows the procedure followed by X25 in the setting up of a virtual circuit data call through the PSS. The calling terminal sends a call-request packet CR, which contains the address of the destination terminal, into the network. The packet also contains a logical-channel number which is used to identify the particular virtual channel being used. At the destination PSE this CR packet is converted into an incoming-call packet CN and this is passed on to the destination terminal. The CN packet contains the address of the calling terminal and a logical-channel number; the latter is used by the called terminal during message transfer. When the CN packet arrives at the destination terminal that terminal can either accept or reject the call. If it accepts the call the terminal responds with a call-accepted packet CA which contains both the logical-channel number and a call-accepted indicator. This CA packet is changed by the PSE nearest to the calling terminal into a call-connected packet CC. Once this CC packet arrives at the calling terminal that terminal can commence sending data packets; the communication may be in one direction only or it may be in both directions at the same time, but in the figure it is assumed that transmission is only from the calling terminal to the called terminal. Either terminal can terminate the virtual circuit by sending a clear-request packet CQ which contains both the logical-channel number and a clear signal. At the other end of the virtual circuit this packet is changed into a clear-indication packet CI. As the CQ packet reaches each PSE a clear confirmation packet CF is returned to the preceding PSE, or to the originating terminal. The CI packet also produces a CF packet response and then the virtual circuit has been completely cleared.

The fast select call is a short message (up to 1024 bits in length) that is included within the call-request packet. If required, a short response may be included within the call-accepted packet.

Most packet terminals are able to support packet exchange simultaneously with two, or more, other terminals. X25 deals with

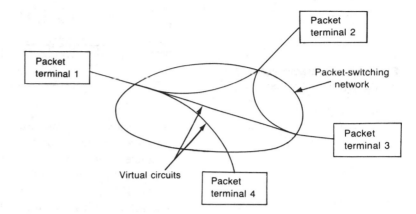

Fig. 11.9 Virtual circuits.

this by employing logical-channel identifiers. A terminal is given a different logical-channel number for each terminal with which it communicates. An example of this is shown by Fig. 11.9 in which one terminal has three simultaneous packet exchanges with other terminals, two terminals have two simultaneous packet exchanges, and the fourth terminal is taking part in only one packet exchange. Each X25 interface has a total of 4096 logical-channel numbers available for use and the choice between them is purely random. At the receiving end of a virtual circuit the choice of the logical-channel number is again made randomly. It is almost certain that the two numbers used at each end of a virtual circuit will be different and the necessary translation between the two logical-channel numbers is carried out by the network.

Packet Terminals and Character Terminals

A *packet terminal* is a computer, or an intelligent terminal, that complies with all three levels of the ITU-T X25 recommendations. Such a terminal is able to format and control packets and to exchange packets with other terminals. Each packet terminal is directly connected to a packet-switching network as shown by Fig. 11.10. A packet terminal operates synchronously at 2.4, 4.8, 9.6 or 48 kbit/s and can communicate with more than one other terminal at the same time. A *character terminal* cannot comply with X25 and so it must be connected to the network via a PAD, that is located at the local PSE. A character terminal can only communicate with one other terminal at a time.

Fig. 11.11 shows two character terminals connected to the PSS;

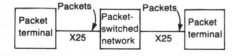

Fig. 11.10 Connection of packet terminals.

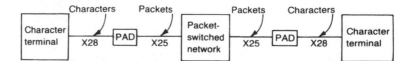

Fig. 11.11 Connection of character terminals.

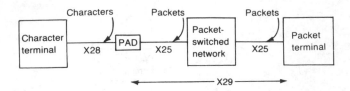

Fig. 11.12 Character-terminal-to-packet-terminal connection.

each terminal communicates with its local PAD using its own character-orientated protocol. The PAD assembles the incoming characters into packets and then passes them to the PSS in accordance with the X25 protocol. The logical operation of a PAD is specified by the ITU-T X3 recommendations and the interaction between the character terminal and the PAD is specified by X28. At the other end of the virtual circuit the packets received by the PAD are disassembled and the recovered message is sent to the receiving terminal one character at a time. When a virtual connection is set up between a character terminal and a packet terminal another ITU-T recommendation, X29, is followed. Fig. 11.12 shows the arrangement. X29 is the protocol used above X25 between a PAD and a packet terminal to present the appearance of a non-synchronous line to the user.

Datalines

The line that connects a terminal to the PSDN is known as a *dataline*. Often it consists of a leased analogue telephone line, with a modem connected at either end, that is operated on a full-duplex synchronous basis. For terminals that have only small amounts of data to send and/or receive the use of a dedicated link to the PSS may not be economic. Such terminals can access the PSS via a dial-up connection over the PSTN; the X25 protocol is then supported at a bit rate of 300/110, 1200/75 or 1200/1200 bit/s. The datalines that are available for use with BT's PSDN service are shown by Table 11.2. The BT Dataline services are the following.

(a) PSS Dataline provides a dedicated circuit between the user's premises and the PSS network that allows most types of computers and terminals to communicate. IBM and IBM compatible equipment will need to use the Multistream service.

Table 11.2

Protocol	Transmission speed (bit/s)	Protocol	Transmission speed (bit/s)
—	300/100	X25	48000
X28	1200/75	BiSynch	2400
X28	1200/1200	BiSynch	9600
X25	2400	SDLC	2400
X25	9600	SDLC	9600

(*b*) PSS Plus is a service for customers who have a requirement for five or more Datalines.

(*c*) PSS Dial provides simple computers and terminals with access to the PSS via the PSTN.

(*d*) PSS Dialplus is a dial-up service which provides access to the PSS via the PSTN with full support for the ITU-T V24 error-correction standard at bit rates of up to 2400 bit/s. All Dialplus transmissions are automatically corrected for any errors using V42.

(*e*) Global Network Services (GNS) Dialplus provides users with access to remote computers and databases at 9600 bit/s. Dialplus and GNS Dialplus are the two most common ways in which customers of BACS connect into the network.

Multistream

Multistream is a service offered by BT that gives its users some extra facilities. There are currently four versions of Multistream known as BPAD, SPAD, EPAD and VPAD.

BPAD

The BPAD Multistream service provides protocol and speed conversions to make it possible for 3270 IBM, and IBM compatible, Bisynch computers and terminals to communicate via the PSS with each other, and also with otherwise incompatible computers and terminals. The basic arrangement of Multistream BPAD is shown by Fig. 11.13. The IBM host computer is connected, via a FEP and a BPAD Dataline, to a BPAD interface at the local packet switching exchange. The BPAD software allows the host to communicate, via the PSS, with other IBM equipment such as a BiSynch control unit and a microcomputer. The host Dataline operates at 9600 bit/s and the terminal Datalines work at either 2400 bit/s or at 9600 bit/s. A link may also be set up with a non-IBM terminal or computer that is connected to the PSS using a normal Dataline.

SPAD

The SPAD Multistream service provides similar facilities as BPAD but for computers and terminals that employ the IBM SNA SDLC protocols. The service allows a number of remote terminals to be connected to a single FEP port via the PSS and, like BPAD, makes it possible for users of IBM, and IBM compatible, equipment to have full use of a nationwide data network without having to set-up their own WAN. The SPAD network is similar to Fig. 11.13 except that SPAD Datalines and SPAD interfaces are employed instead of BPAD.

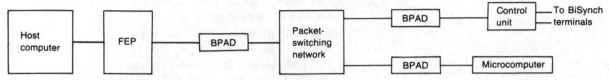

Fig. 11.13 BPAD network.

EPAD

EPAD provides error correction on the dial-up link between a customer's non-synchronous terminal and the PSS network.

VPAD

The VPAD service provides its users with data in the videotex format to enable private viewdata systems to operate nationally via the PSS.

Cell Relay and Frame Relay

X25 packet switching was originally specified to cope with data transmission over the then analogue PSTN. The protocol at each X25 switch checks a received packet for errors, informs the transmitting station of its receipt and then sends the packet on to the next node. The time taken for the transfer of data through the switches causes the effective data transmission rate of a connection to be less than the capacity of the high-speed interconnecting circuits.

Two modern developments of X25 packet switching are known as *cell relay* and *frame relay*.

The main features of cell relay and frame are as follows:

(a) Connections are not error-controlled since it is assumed that the communicating terminals will be able to detect and correct any error that may occur. However, since frame relay has been designed for use over digital circuits which have, typically, an error rate of 10^{-8}, few errors occur. Many of the network delays and protocol overheads associated with X25 are reduced by the use of frame relay. The absence of link error checking increases the data transmission rate through the network since delays at each node (largely caused by error checking in X25 systems) are considerably reduced.

(b) Multiple sessions can be handled and flow control is not provided.

(c) Cell relay uses packets of fixed length and allows virtual circuits to be set up over a single network, while frame relay data packets are of variable length. Frame relay's frame structure, known as LAP-D, is very similar to that of X25 except for the header. The frame structure is shown by Fig. 11.14.

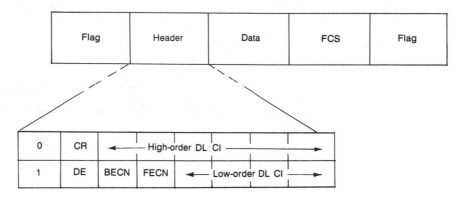

Fig. 11.14 Frame structure of frame relay message.

(*d*) Hardware switching is employed and a wide range of bit rates for all kinds of data traffic can be supported.

Radio Data Systems

There are a number of two-way radio packet switching data networks in the UK. Two of the more commonly used are known as Vodata and Paknet. Both of these systems transfer data, via a mobile cellular telephone and a modem, in small packets. The basic arrangement of the Paknet system is shown by Fig. 11.15. Data terminals/computers are connected to a radio PAD: this performs a similar function to the PAD used with the PSDN but incorporates a VHF radio FSK transceiver. Each radio PAD has two ports which can be used for

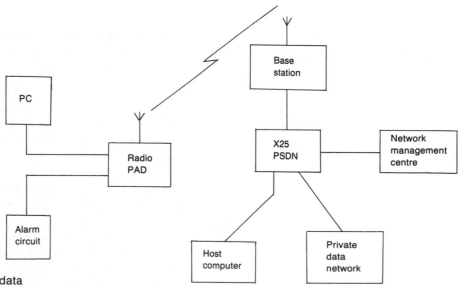

Fig. 11.15 Radio packet data system.

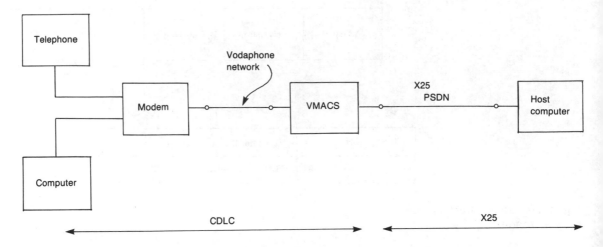

Fig. 11.16 Vodata packet data system.

two quite independent terminals. At the other end of the VHF radio link is the base station which, in turn, is connected to the PSDN. Each base station has the capacity to provide service to several radio PADs at the same time. The base station contains both a radio transceiver and an X25 interface equipment. The frequencies allocated to the Paknet system are in the band 136 to 174 MHz, and the bandwidth is 12.5 kHz.

The Vodata system is also employed in conjunction with the X25 PSDN, and Fig. 11.16 shows the basic set-up of a connection between a mobile and a host computer. The operation of the circuit and the function of the VMACS are very similar to the system described on page 194, which uses the PSTN to gain access to the host computer.

Relative Merits of Message Switching and Packet Switching

Message-switching and packet-switching networks are both used by business, commerce, etc., and a number of private networks have been installed. The relative merits of the two systems are as follows.

(a) A message-switching system transmits the messages as a whole but the transmissions are one-way only and sometimes considerable delays are experienced.

(b) A packet-switching system provides real-time interaction between two terminals since packets can be sent in both directions at the same time.

(c) In a packet-switching system all the incoming data arrives in exactly the same format, this is not always true of a message-switching system.

(d) The large numbers of low-capacity links used in a message-switching system are replaced by much fewer high-capacity links in a packet-switching system. This is possible because of the use of time-division multiplexing on the links in a packet-switching system.

12 OSI and ISDN

Two internationally supported developments have been proposed and are now beginning to be implemented throughout much of the world. These developments, known respectively as the Open Systems Interconnection (OSI), and the Integrated Services Digital Network (ISDN), will have a considerable impact on the ways in which data communication is provided in the future, and so a brief introduction to both developments is given in this chapter.

The OSI is being introduced by the International Standards Organisation (ISO) and its aim is to ensure that every computer/terminal connected to a network is able to communicate with any other computer/terminal that is connected to either the same network or to any other linked network. This is known as an *open system* and its basic concept is illustrated by Fig. 12.1.

In this figure four computers, made by four different manufacturers, are connected to two different networks that are linked together by a gateway. The two networks are also supplied by different manufacturers. Each link between a computer and a network may be an analogue line with modems, or a digital line with network terminating units. Errors can occur on any such link and to detect, and correct, such errors a link protocol is necessary. Errors can also occur within a network and so another protocol must be used across each network to guard against those errors. Each of the link controls has essentially only the one task, this is to ensure that each block of data that is transmitted by a computer into a link arrives at the other computer without error. A further requirement is, of course, that a block of data entering a network is switched through the network to its correct destination. A fourth level of protocol is then needed to ensure that blocks of data are transmitted without error from one computer to the other through as many networks as is necessary; this will be an end-to-end protocol. For complete compatibility to be achieved a set of standard communication protocols is necessary. In

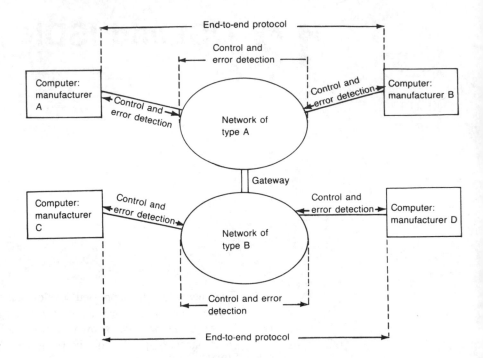

Fig. 12.1 Open system.

the past it became increasingly obvious to the computer and data communication industries that the different communication protocols used by various manufacturers would make it extremely difficult, if not impossible, to achieve the desired compatibility between computers and networks. The development of OSI was therefore supported by most, if not all, computer manufacturers and telecommunication authorities.

The OSI model has identified all of the functions required for an open system and has divided them into seven layers. The layers describe, in a way that does not specify how the implementation will be carried out, the functions of each layer. Thus the OSI model is a set of standardized protocols which will provide for communication between computers and/or terminals regardless of their manufacturer. The ISO model is specified by ISO 7498 and by ITU-T X200.

The public telecommunications systems provided in the UK by BT are the public switched telephone system (PSTN) and the public switched data system (PSDN). The PSTN was originally designed for the transmission of speech signals but it is now also employed for data communication. The core telephone network, which consists of trunk circuit and trunk switching exchanges, is a digital system that uses pulse code modulation. Local telephone exchanges are being converted to digital operation as rapidly as possible. At present the access network remains predominantly analogue but eventually it too will become digitally operated; the same telephone line will still join the customer to the local exchange but each telephone will incorporate a circuit known as a *codec* that will convert analogue speech into digital signals and vice versa. As the local line network is converted to digital operation the ISDN can be progressively introduced. The ISDN is a synchronized hierarchy of digital transmission and switching systems that give compatible transmission of data, video and voice signals.

Open Systems Interconnection

The reference model of the ISO's Open Systems Interconnection (OSI) scheme is shown in Fig. 12.2. The various communication functions have been placed into seven vertical layers; each layer provides a service to the next higher layer. The fundamental requirements of each layer in the scheme are specified in a way that is independent of the manner in which the requirements are implemented. This has been done so that each manufacturer can decide how best to satisfy the requirements. The model does specify the external behaviour of a system, i.e. the communications protocols, since this is what allows the interconnection of equipments made by different manufacturers. The OSI model describes the functions that every station must perform before a message can be transmitted and received. When a message is to be passed from one terminal to another it passes downwards through the layers at the transmitting terminal, from the top to the bottom, with each layer adding its own header to the message. At the data link layer the message is placed into a frame that has both a header and a trailer in a way specified by the protocol of the network, e.g. HDLC. This is shown by Fig. 12.3. The message is then transmitted across the network to the receiving terminal; here the message travels upwards from the bottom layer to the top layer. At each layer the message is stripped of its header and the layer acts on the enclosed protocol before it passes the message to the next higher layer. The OSI model defines a framework in which protocols may be fitted but it does not describe the actual protocols themselves. It defines the functions of each layer and the services that each layer performs for the next higher layer. The seven layers are as follows.

Physical Layer

The physical layer provides the means for bits to be transmitted across a physical communication path. The layer defines the electrical and

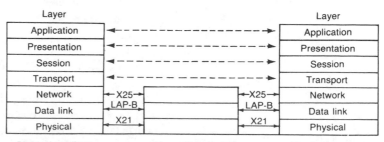

Fig. 12.2 The OSI scheme.

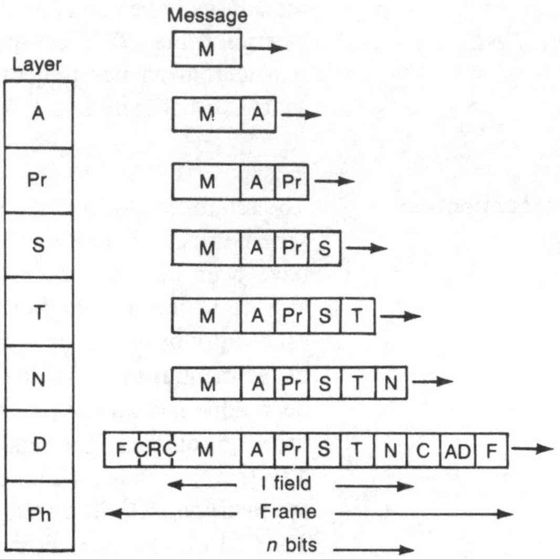

Fig. 12.3 Showing how a message gathers headers as it passes downwards through the OSI layers. F = flag; AD = address field; C = control field; M = message; CRC = error-checking mechanism.

mechanical requirements of the communication system. It deals with most of the hardware problems, e.g. the type of cable to be used, the type of connectors to be employed, the bit rate(s) and signal levels to be used. It also defines the topology of the network, e.g. bus or ring. Examples of the physical layer are ITU-T V24, EIA 232E and EIA 449.

Data-link Layer

The data-link layer provides the transport of bits of data grouped into 'frames' from one point to another across the physical link. It provides a protocol, with addressing and error-checking facilities added, that determines how a terminal can gain, and later give up, access to a network. The main examples of this layer are the HDLC and LAP-B protocols. The protocols deal with such matters as the framing of data blocks, synchronization, error detection, and flow control. The error detection is only applied to each link separately; if there is just the one link in a connection the higher layer will inherit an error-free connection but if there are two, or more, links in the connection overall error detection must also be exercised at a higher layer.

Network Layer

The network layer addresses and routes data through a network or a series of networks. It controls the routeing and switching of messages in a manner that is independent of the actual network in use. It is responsible for the setting-up, maintenance, and clearing-down of a circuit. The layer provides the protocols for communication between different networks and so it is important in both dial-up and gateway applications. An example of this layer is the ITU-T X25 recommendations.

Transport Layer

The transport layer deals with the choice of which type of network is to be used for a particular communication. It is the lowest layer in which a protocol operates end-to-end to give reliable, transparent, transfer of data between two terminals. It is the layer that ensures that a message, probably consisting of several frames, reaches its correct destination; it does this by defining the addressing of the destination and by specifying how connections, both within and between networks are set up and cleared down. The layer is responsible for dividing a message into the particular frame format that is required by a network and for its reassembly at the receiving terminal. The layer also ensures that the frames arrive at the destination in their correct order, and provides end-to-end error correction and data flow control. Error correction is achieved by the re-transmission of faulty messages.

Session Layer

The session layer handles logon and logoff procedures and then it establishes, and later clears down, the connection between two terminals. It controls the transfer of messages over the network and overall error detection. The layer controls how a message starts and finishes, whether or not a message is to be acknowledged, and whether the link is to be operated on a half-duplex, or a full-duplex, basis. Lastly, the layer maps logical names on to physical addresses and so allows software to be written that will run on any kind of network.

Presentation Layer

The presentation layer makes sure that the message received by a terminal can be understood by that terminal. This means that it deals with the selection and the structure of codes and with any necessary translation between different formats, codes, languages, and transmission speeds. It deals with character sets, page layouts and encryption.

Application Layer

The concern of the application layer is the setting-up of the overall connection of one terminal to another and later the later clearing down of that connection. It provides the interface between the communications interface and the applications process. The application layer provides an actual service to the user, such as the sending of a bill or a document or the transfer of a file. It supplies standards for such applications as electronic mail, EFTPoS and ATM. The layer makes available such services as passwords and any procedures that a particular industry or business may wish to employ. The application layer is the only layer that is not transparent to the user. In a LAN 802.3, 802.4 and 802.5 are all part of the physical layer and 802.2 is in the data link layer. In networking of LANs a repeater is at the physical layer, a bridge is at the data link layer, a router is at the network layer, and a gateway is at the application layer.

Integrated Services Digital Network

The integrated digital transmission and switching network that is being set up in the UK is known as the *Integrated Digital Network* (IDN). It will allow 64 kbits/s connections to be set up on a call-by-call basis using circuit switching.

At present the core network is digital and so are the vast majority of the local telephone exchanges. When the local line network has been digitalized it will be known as the *Integrated Digital Access* (IDA). The combination of both the IDN and the IDA will be used to offer customers a multi-function end-to-end digital service that will be known as the *Integrated Services Digital Network* (ISDN). Digital exchanges are interconnected by a digital switching network and each exchange is connected to its customers by a digital local line. Each local line terminates on a network terminating unit (NTE) whose function is to convert the line code into the code used by the interface standard EIA 232E.

BT offers two ISDN services, a basic rate service, and a primary rate service.

The basic rate service is intended for small business and home users and its service is known as either ISDN-2 or as 2B + D. 2B stands for two bearer circuits which are each allocated a separate number on the local telephone exchange, and D stands for a data circuit. The 2B + D service is shown in Fig. 12.4; it consists of two 64 kbits/s B channels and one 16 bits/s D channel, a total of 144 kbits/s. The 64 kbits/s channels may be used to carry either speech or data signals while the functions of the D channel are to carry signalling information, to control the setting up and clearing down of calls set up on the two B channels, and to provide end-to-end synchronization. The D channel can also be used to transmit data using X25 packet switching. Figure 12.5 shows how a connection might be set up between two PCs over a link in the ISDN. Because ISDN lines are installed with two bearer circuits and a common control line their individual performance may be merged to give a single 128 kbits/s circuit. This merging process is known as *reverse multiplexing*. The

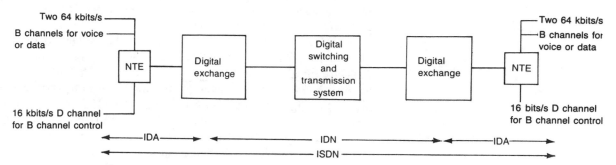

Fig. 12.4 Basic ISDN scheme.

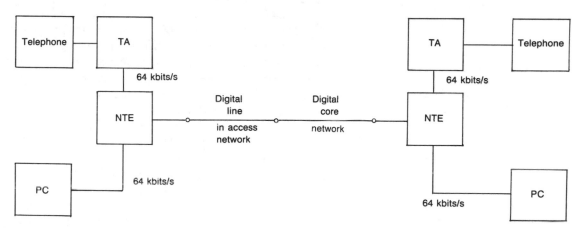

Fig. 12.5 ISDN connection between two PCs.

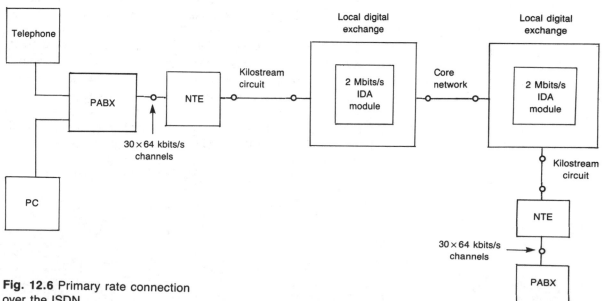

Fig. 12.6 Primary rate connection over the ISDN.

two B channels are each allocated a separate number on the local telephone exchange.

The primary rate service is known as either ISDN-30 or as 30B + D; it provides the user with thirty 64 kbits/s speech or data channels plus a separate 16 kbits/s control channel. A primary rate circuit is intended for use by large businesses and to be connected directly to a digital PABX as shown by Fig. 12.6. The PABX is then used to switch lower bit rate circuits to their destinations, which may be either a telephone or a data terminal or computer.

Where Mercury has provided local lines into a customer's premises they are digital circuits, and hence a direct ISDN service is available. Mercury also provide an indirect ISDN service, known as ISDN 132, that employs a BT ISDN circuit from the local telephone exchange to the home/office.

If a terminal supports the ISDN communication protocol it may be connected directly to the NTE; such a terminal might be a PC (fitted with an ISDN card), a mainframe computer, or a LAN router. If a terminal does not support the ISDN communication protocol then a *terminal adaptor* is required. The terminal adaptor is fitted between the terminal and the NTE and its function is to convert the communication protocol supported by the non-ISDN equipment into the protocol required by ISDN-2.

The use of the ISDN for the transmission of data has some advantages over the use of modems and the PSTN. The first advantage is that the bit rate made available is much higher at 64 kbits/s, and second is the much faster set-up time. On the other hand, most PCs are unable to transmit data at 64 kbits/s, since a fast UART like the 16550 is required.

The services that will be offered to each ISDN customer include telephony, high-speed data and video. Each customer will be provided with a switched data service that will not require the use of a modem and that can be employed for such applications as group IV FAX, Teletext, Datel, videotex and slow-scan television. Access to the PSS will be given to all ISDN customers via the digital network. Kilostream circuits will provide a multi-line IDA (30B + D) service that will give a customer a 30-channel PCM system between his premises and the local digital exchange. At the customer's premises the system will terminate on a digital PABX and at the exchange on a 2 Mbit/s IDA line module.

Exercises

Chapter 1

1.1 What is meant by the terms synchronous and non-synchronous when applied to a data circuit? Explain how synchronous transmission gives rise to a more efficient utilization of the transmission medium than does non-synchronous transmission. Give an example of each waveform.

1.2 What is meant by the following terms used in synchronous data transmission: (a) bit synchronism, (b) character synchronism, and (c) message synchronism?

1.3 A non-synchronous data signal is transmitted at 1200 bit/s. The receive clock is 5% slower than the clock in the transmitter. How many bits would be accurately received before an error occurs? Assume that both the clocks start at the same time.

1.4 Discuss the merits of having the transmitter clock in (a) the data terminal, (b) the interface circuit or (c) the modem.

1.5 The message 42H, 16H, 33H is transmitted over a data link, 42H first. Due to a loss of synchronization the first data bit transmitted is lost. Determine the received message.

1.6 The characters Ab are sent over a synchronous circuit. Explain (a) how false synchronization could occur, (b) how it is avoided.

1.7 A block of 240 ASCII characters is sent over (a) a synchronous, (b) a non-synchronous circuit. The synchronous signal is preceded by two SYN characters. Calculate the percentage increase in transmission efficiency of (a) over (b). Assume one parity bit per character in each case.

1.8 The signal level received at the end of a data circuit is equal to 10 mW. Calculate the signal-to-noise ratio at the receiver if the received noise level is 7 μW.

1.9 (a) What is meant by (i) bias distortion and (ii) bit jitter? (b) In a received data signal the 1 pulses are 10% longer than the 0 pulses. Calculate the percentage bias distortion. (c) A digital data signal has a bit jitter of 80 μs. Calculate the percentage jitter if the bit rate is 4800 bits/s.

Chapter 2

2.1 The data signal produced by a terminal uses the seven-bit ASCII code plus an even-parity bit. Non-synchronous transmission is used with one stop bit. Calculate the bit rate if 60 characters per second are transmitted. The signal is applied to an FSK modem. State the frequencies that are used to represent binary 1 and 0 and draw the FSK waveform.

2.2 Explain the meanings of the terms isochronous and anisochronous as applied to the transmission of a data waveform. Draw an anisochronous waveform in which each bit has a time duration of 833.3 μs and calculate (a) the baud speed, (b) the bit rate, and (c) the maximum and minimum component frequencies of the waveform.

2.3 Explain the principles of differential phase shift modulation using the ITU-T standard values to illustrate your answer. What is meant by the term dibit? Explain why the use of dibits reduces the bandwidth necessary to transmit a data signal. Calculate the line baud speed if the bit rate is 2400 bit/s.

2.4 Alphanumeric characters are transmitted as seven-bit ASCII words with a single parity bit added. The bit rate is 9.6 kbit/s. (a) Calculate the number of characters transmitted per second. (b) If a typical page of text contains 500 words having an average length of 5 characters per word and one space between words, calculate how long it will take to transmit one page.

2.5 The 1200 bit/s data signal 10110011 is applied to an FSK modulator. Draw the waveform of the resulting FSK signal and indicate the frequencies used.

2.6 (a) Calculate the maximum fundamental frequency of a 4.8 kbit/s data signal. Calculate the line baud speed if (b) dibits and (c) tribits are used.

2.7 (a) Explain the difference between the phase velocity and the group velocity of a signal. (b) What is meant by group delay and by group-delay distortion? Explain the effect they may have upon a data signal.

2.8 Explain the principles of quadrature amplitude

modulation and explain why it is used at the higher bit rates. State the ITU-T V systems that use this form of modulation and for each one quote the line baud speed.

Chapter 3

3.1 State the bit rate, the modulation method and the line baud speed for each of the following modems: (a) V22, (b) V23, (c) V29 and (d) V32. Discuss the relative advantages of using the PSTN or leasing a private circuit for each of the above.

3.2 What is the function of a modem? Draw the basic block diagram of a modem giving outline details of the transmitting and the receiving sections. Explain the importance of the ITU-T recommendation V24 and give three examples of the circuits contained in it.

3.3 The specification of a modem includes the following: 9600/7200/4800 bit/s synchronous operation; point-to-point, four-wire, full-duplex over leased circuits; dial back-up for PSTN. Explain the meaning of each term in the specification.

3.4 The commercial-quality speech bandwidth is 300 Hz to 3400 Hz. How much of this can be used by data signals sent over the PSTN? Why can DPSK signals use more of this bandwidth than can FSK signals?

3.5 (a) Explain what is meant by a multi-port or multi-plexed modem and discuss the use of such an equipment. (b) What is meant by a modem-sharing unit and when and where might it be employed?

3.6 What is meant by 'downward compatibility' of a modem? Quote typical figures. How do two modems negotiate the 'fallback'?

3.7 Explain the techniques which make it possible for a modem to operate at a bit rate of 9600 bits/s or more over a PSTN connection to provide full-duplex service. Use headings of modulation method, line conditioning and echo cancellation.

3.8 Expalin briefly what is meant by 'Trellis encoding'. With which ITU-T V standards is it used? Why is it not used with (a) FSK and (b) DPSK modulation?

Chapter 4

4.1 Explain the function of each pin of the MC6821 PIA.
4.2 Explain the function of each pin of the MC6850 ACIA.
4.3 List the EIA interfaces that are used in data circuits. Which EIA specifications relate to the electrical characteristics of interface circuits, and to which ITU-T recommendations do they correspond? Over which voltage ranges is a control circuit turned ON for (a) EIA 232, and (b) EIA 422?

4.4 Explain why there is a need for modem/terminal interface standards. List the most important of the ITU-T V24 interface circuits and explain the function of each of them.

4.5 Explain the operation of the Centronics parallel interface.

4.6 What is meant by a null modem? When and where would such a circuit be employed? Draw the connections of a null modem and briefly say how it overcomes the problem of a V24 interface between two terminals, or between two computers.

4.7 Explain clearly the difference between a peripheral interface adaptor and an asynchronous communication interface adaptor. Draw block diagrams to show how each IC is used. Both devices are given a variety of names by other manufacturers; for each give two alternative names.

4.8 Write a segment of a program that would configure ports PA_0 to PA_4 as inputs and the remaining PA, and all the PB ports as outputs. The address of port A is $4050/1 and of port B is $4502/3. The contents of memory locations $85 to $87 are to be sent to the peripherals connected to the PA output ports. Write a possible program.

4.9 Explain the functions of the (a) control register, and (b) data direction register of a 6821 PIA.

4.10 The following questions all refer to a 6850 ACIA. (a) To which register does the data sent by the microprocessor go? (b) What types of errors are detected when receiving data? (c) To which microprocessor address line should the RS pin be connected? (d) How is the IC reset?

Chapter 5

5.1 Four 4.8 kbit/s data terminals are to be linked to a distant computer using TDM. Calculate the required bit rate capacity of the bearer circuit.

5.2 Explain clearly the difference between (a) a multiplexer and a concentrator and (b) a data concentrator and a line concentrator. Why may a statistical multiplexer be regarded as either a multiplexer or a concentrator?

5.3 A number of 1200 bit/s data circuits are to be operated over a 4800 bit/s bearer circuit using time-division multiplex. If bit stripping is used calculate the number of circuits that can be accommodated.

5.4 Fig Q.1 shows a data circuit. If each character contains

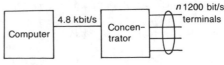

Fig. Q.1

10 bits, the average message contains 60 characters, and on average one message is sent by each terminal every 60 seconds, calculate the number n of channels. Assume that the concentrator must not be more than 50% utilized.

5.5 Explain the difference between a multiplexer and a statistical multiplexer. Illustrate your answer by considering eight 4.8 kbit/s data circuits routed over a single 9.6 kbit/s bearer circuit.

5.6 Why is the increase in efficiency gained by the use of bit stripping in a multiplexer less than is calculated? What is meant by a multiplexer being transparent? Why is character interleaving generally used for multiplexing non-synchronous circuits but not for synchronous circuits? Why will a 1 ms burst of noise affect a bit-interleaved multiplexed system more than a character-interleaved multiplexed system?

5.7 Explain the principle of operation of a terminal that accepts data in parallel form and transmits it over the PSTN. A terminal has a speed of 4800 bit/s and is used to transmit the contents of 80 files. If each file contains on average 250 lines with 50 characters/line calculate the transmission time. State any assumptions made.

5.8 A computer is linked via a multiplexer to a number n of low-speed terminals. The terminals operate at 120 characters/s, the average message length is 700 characters, on average one message is transmitted every 120 seconds and there are 10 bits/character. Calculate the number of terminals that can be connected to the system if the bearer circuit is able to operate at up to 4800 bit/s. State any assumptions made.

5.9 What is the difference between a multiplexer, a statistical multiplexer, and a concentrator? Illustrate your answer by supposing that the inputs are four 4800 bit/s, one 9600 bit/s. and 16×1200 bit/s channels and suggest a suitable maximum bit rate for the bearer circuit.

5.10 (*a*) Three 4800 bit/s data channels are applied to the input of a multiplexer. Calculate the bit rate on the output line. (*b*) Six 4800 bit/s data channels are connected to the input of a concentrator. Calculate the bit rate on each of the three identical output circuits.

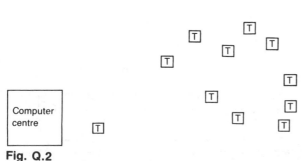

Fig. Q.2

Chapter 6

6.1 A computer centre is to be connected to a number of data terminals that are situated around the centre in the way shown by Fig. Q.2. Draw two networks that could be used, one using modems and analogue lines the other using Kilostream circuits.

6.2 Explain the functions carried out by a front-end processor in a data network and say why these tasks are not left for the main computer to carry out. If the network contains several computers the need to provide several FEPs may prove expensive. What other alternative is there and what extra facilities can then be obtained?

6.3 A data terminal installed at a point remote from a main computer consists of two equipments, one of which operates at 2400 bit/s and the other at 4800 bit/s. Draw diagrams of two suitable links if (*a*) both equipments work at the same time, and (*b*) only one equipment works at a time. The line to the computer has a maximum bit rate of 4800 bit/s.

6.4 (*a*) With the aid of a diagram explain the operation of a multi-drop data circuit and explain the meaning of the term 'polling'. (*b*) What is meant by the terms half-, and full-duplex, and two-, and four-wire presented circuits? Why is it sometimes necessary to have a four-wire presented circuit if full-duplex operation is wanted? How can high-speed modems provide full duplex operation over a PSTN connection?

6.5 Explain the advantages of a digital data network over an analogue data network. Use the headings: (*a*) use of digital technology, (*b*) time-division multiplexing, and (*c*) control signals.

6.6 The equipment at one site consists of a 9.6 kbit/s CAD terminal, two multiplexers that output data at 9.6 kbit/s and 4.8 kbit/s, respectively, two 9.6 kbit/s computer ports, and the 16 kbit/s output of a digital PABX. This site is to be connected by a Kilostream circuit to another site. Here there are three 9.6 kbit/s terminals, one 9.6 kbit/s multiplexer, one 4.8 kbit/s terminal and a 16 kbit/s digital PABX. Draw a diagram of a suitable network.

6.7 A Megastream circuit is used to interconnect two computer sites. At each site the equipment consists of three 128 kbit/s multiplexers, one 64 kbit/s statistical multiplexer, a 64 kbit/s digital PABX, and five terminals that operate at 512, 64, 256 and 128 kbit/s. Draw a possible network.

6.8 List the ITU-T V services. For each service state (*a*) the bit rates and the modulation system, (*b*) whether it is synchronous or non-synchronous, and (*c*), whether half-, and/or full-duplex facilities are provided.

6.9 Explain the facilities offered by the BT services Kilostream, Megastream, Satstream, Packetstream, Multistream and Netstream.

6.10 Explain how data is transmitted over (*a*) a Kilostream circuit, and (*b*) a Megastream circuit. Show how the transmission speed of each circuit can be used to carry a large number of lower-speed channels.

Chapter 7

7.1 Explain what is meant by a value added network and list five VANs. Give a brief description of each VAN listed.
7.2 What is meant by electronic mail and why is it playing an ever-increasing role in modern offices? Explain how an electronic mail system works.
7.3 What is an EPoS system? Discuss the use of such systems in supermarkets and department stores and explain how considerable economies are obtained. Sketch a possible network for connecting the in-store EPoS tills to computers at (*a*) head office and (*b*) the warehouse.
7.4 Commerce and business are making increasing use of electronic fund transfer. Discuss the advantages to (*a*) businesses and (*b*) banks of EFT.
7.5 Explain the differences between the use of the FAX system and electronic mail. Discuss their relative merits.
7.6 A customer of a high-street bank inserts his cheque card into a cash dispenser and when requested enters his PIN number. Outline the sequence of events which takes place before a cash request is either rejected or implemented. Why is it not satisfactory for the dispenser to be linked by a PSTN connection to the bank's computer?

Chapter 8

8.1 A 'ping-pong' protocol is used on some 4.8 kbit/s and 9.6 kbit/s circuits to simulate full-duplex operation. The protocol automatically allocates the direction of transmission on the line in proportion to the amount of data being sent. Explain how this technique can make a half-duplex line appear to be full-duplex.
8.2 The format of an X25 frame consists of an eight-bit flag, an eight-bit address, an eight-bit control, *n* data bits, a 16-bit FCS and, lastly, another eight-bit flag. Explain the purpose of each part of the frame.
8.3 The specification for a data equipment includes the following: character length five—eight bits, bit stuffing/stripping, optional PAD opening, generation and detection of flag, abort and idle bit patterns, and CRC generation/checking. Explain the meaning of each of these terms.
8.4 The protocols employed in data systems fall into one of three main classes. State these three classes and give one protocol that is an example of each. For each example given draw the format of a data frame.

8.5 A long message is split into four blocks for transmission. For both the BiSynch and the HDLC protocols show the frame sequences that would be transmitted if the third frame has an initial error but is accurate after a retransmission.
8.6 What does the term 'transparent' mean when used in data communication? How are control characters made transparent in the BiSynch protocol?
8.7 Explain why the use of a protocol is necessary for data to be transferred from one terminal to another. Describe and compare the principal features of the BiSynch and HDLC protocols.

Chapter 9

9.1 Explain the principle of the Hamming bit method of forward error correction. Why is this method of error correction not often used? Determine the Hamming bits in the bit stream 11001100011.
9.2 The data word 1001011 is applied to the parity generating circuit given in Fig. 9.1. Demonstrate that the even-parity bit is a 0. Show that if the second bit is inverted the parity bit will become a 1. How can the circuit be used at the receiver to detect a single bit error? Confirm that the circuit is unable to detect two bits in error.
9.3 Explain the meanings of (*a*) FCS, (*b*) BCC and (*c*) CRC. With which protocols are (*b*) and (*c*) used? What is meant by the terms 'longitudinal redundancy check' and 'vertical redundancy check'?
9.4 For the ASCII coded block of data shown (*a*) decipher the message, (*b*) determine the BCC.

```
1 0 0 0 1 1 1
1 0 0 1 1 1 1
1 0 0 1 1 1 1
1 0 0 0 1 0 0
0 1 0 0 0 0 0
1 0 0 1 1 1 0
1 0 0 1 0 0 1
1 0 0 0 1 1 1
1 0 0 1 0 0 0
1 0 1 0 1 0 0
0 0 0 0 0 1 1
```

9.5 The data received over a line is

$$10001 \quad 01100 \quad 11000 \quad 10000 \quad 10001$$

where the least significant bit, and the least significant word, are on the right. In the first four words the parity bit is in the most significant position and the fifth word is all parity bits. Determine (*a*) the parity system used, (*b*) the bit in error, (*c*) the received hex number, and (*d*) the correct number.

9.6 (*a*) What is meant by odd and even parity? (*b*) The message 'Sixty six trombones' is to be sent over a data circuit using BiSynch. Determine the BCC. Assume the message to be preceded by SYN SYN STX and followed by ETX. (*c*) At the receiver the BCC is in error in its lsb and the ninth character has a polarity error. What is the received message?

9.7 Explain the principle of a CRC. The data 1110010010 is to have a four-bit CRC calculated from it. Choose a suitable divisor and then calculate the CRC.

9.8 Design and draw the circuit that will generate the CRC calculated in 9.7.

9.9 Explain the principle of error detection and correction of a block of ASCII data. Illustrate your answer by showing a block with one error. Explain the relative merits of correcting detected errors by (*a*) retransmitting the faulty block of data, and (*b*) inverting the bit(s) in error.

Chapter 10

10.1 Distinguish between local area networks, wide area networks, metropolitan area networks and value-added networks.

10.2 What is meant by the term local area network and what are the main advantages to be gained by the use of one? Outline the main features of two different kinds of LAN.

10.3 What are the three main kinds of LAN and what network topology does each of them usually employ? Compare and contrast the main features of each system.

10.4 Explain the operation of a token-passing LAN. How is a station prevented from monopolizing the system? How can a bus LAN be operated logically as a ring? What happens when one of the stations connected to a LAN is temporarily non-operative?

10.5 An Ethernet LAN uses the CSMA/CD technique. What do these initials stand for? A LAN is built up from a number of segments; what does each segment consist of and how are the segments connected together? Briefly outline the operation of the LAN.

10.6 Distinguish between the terms 'baseband' and 'broadband' when applied to a LAN. Why is it likely that broadband systems may have a limited future? What are the main applications of the two systems?

10.7 What is meant by a 'server' and why are they required in a LAN? What are the differences between a terminal server, a printer server, and a disk server? A large LAN may often employ a dedicated hardware server while a LAN of PCs will probably employ a PC, running suitable software, as a server. Discuss the relative merits of each type of server.

10.8 What is meant by a peer-to-peer network? When would one be used instead of a LAN with dedicated servers? List the relative merits of peer-to-peer and NOS LANs.

10.9 List the three kinds of equipment which may be used to interconnect two, or more, LANs. Under each heading outline the functions of the device and state when it would be employed.

What is meant by 'terminal emulation' and when and why would it be used?

Chapter 11

11.1 Explain why there is an application for packet switched networks. Draw the block diagram of a typical system and explain how it works. Give the format of a packet frame.

11.2 Draw the diagram of a message-switching system and explain its operation. Why cannot such a system provide real-time operation? List the advantages of a message-switching system.

11.3 Explain the meanings of the following terms that are used in conjunction with a packet-switching system: (*a*) Dataline, (*b*) packet terminal, (*c*) character terminal, and (*d*) PAD.

11.4 Draw the diagram of a packet-switching network and explain its operation if the routeing of packets through the network is done on (*a*) a per-packet, and (*b*) a per-call basis. Which of these two methods is generally employed and which protocol does it follow?

11.5 Draw a diagram to show the format of a packet in a packet-switching system. What protocols do the packet header and the frame header follow? Explain the function of the address field and state the possible addresses. How does the control field indicate to the receiver whether a frame is an information frame, a supervisory frame, or an un-numbered frame?

11.6 (*a*) What is the BT's Multistream service and what facilities does it offer? (*b*) Briefly explain the ITU-T V42 standard. (*c*) Briefly explain the LAP-B protocol. How are (*b*) and (*c*) connected with (*a*)?

11.7 Make a list of the relative advantages and disadvantages of (*a*) circuit switching, (*b*) message switching, and (*c*) packet switching.

11.8 Discuss the reasons why increasing use is being made of radio packet data systems. With the aid of a block diagram explain the operation of a typical system.

Answers to Numerical Exercises

Chapter 1

1.1 5% of 1200 = 60 bits/s. Hence x = 1/1200 and y = 1/1140.
n/1200 + 1/2400 = n/1140
n = $1/(2400 \times 4.386 \times 10^{-5})$ = 9.5.
Therefore 9 bits. (*Ans.*)

1.5 Sent signal = 33H 16H 42H, (42II first)
            = 0011, 0011, 0001, 0110, 0100, 0010
    If the first bit sent is lost the received signal is
            0001, 1001, 1000, 1011, 0010, 0001
            = 19H, 8BH, 21H (*Ans.*)

1.7 (*a*) Bits = $240 \times 8 + 2 \times 8 = 1936$
(*b*) Bits = $240 \times 10 = 2400$
Increase in transmission efficiency
    = 100[(2400 − 1936)/1936]
    = 24% (*Ans.*)

1.8 Signal-to noise ratio
    = $10 \log_{10}[(10 \times 10^{-3})/(7 \times 10^{-6})]$
    = 31.6 dB (*Ans.*)

1.9 % jitter = $80 \times 10^{-6} \times 4800 \times 100$ = 38.4%
                                                    (*Ans.*)

Chapter 2

2.1 Bit rate = $60 \times 10 = 600$ bits/s (*Ans.*)
2.4 (*a*) 8 bits/s character so characters/s = 9600/8 = 1200
                                                    (*Ans.*)
(*b*) If space = character then average characters per word = 6 and characters per page = 3000.
Time per page = 3000/1200 = 25. s (*Ans.*)
2.6 (*a*) f = 4800/2 = 2400 Hz (*Ans.*)
(*b*) Line speed = 4800/2 = 2400 bauds (*Ans.*)
(*c*) Line speed = 4800/3 = 1600 bauds (*Ans.*)

Chapter 5

5.1 Bit rate = $4 \times 4.8 = 19.2$ kbits/s (*Ans.*)

5.3 Number of circuits = $(4800/1200) \times 10/8 = 5$ (*Ans.*)
5.4 On average each message contains 600 bits and, at 1200 bits/s, takes 0.5 s to transmit. There is a message every 60 s on average so there is a maximum of 60/0.5 = 120 channels. But the concentrator is only 50% utilized, hence number of channels = 60 (*Ans*)
5.7 Characters sent = $80 \times 250 \times 50 = 1 \times 10^6$
Assuming synchronous transmission, time taken
    = $(7 \times 10^6)/4800$
    = 1458.3 s
    = 24.3 minutes (*Ans.*)
5.8 Terminal bit rate = $120 \times 10 = 1200$ bits/s.
If terminals sent data continuously, n = 4800/1200 = 4. On average terminals send 7000 bits every 120 seconds and each message takes 7000/1200 = 5.833 s.
Therefore, maximum number of terminals
    = $4 \times (120/5.833)$
    = 20 (*Ans.*)
5.10 (*a*) $4800 \times 3 = 1440$ bits/s (*Ans.*)
(*b*) 4800 bits/s (*Ans.*)

Chapter 9

9.1 Bit stream is X1100110X001X1XX.
1 bits occur in positions 3, 5, 10, 11, 14 and 15.

Table 1

15	1	1	1	1
14	1	1	1	0
11	1	0	1	1
10	1	0	1	0
5	0	1	0	1
3	0	0	1	1
	0	1	1	0	= Hamming bits (*Ans.*)

9.4 GOODNIGHT, 11111100
9.6 01011111, SIXTY SIX TROMBONES
9.7 1101, using $X^4 + X^3 + 1$ as divisor

Index